GLOBAL WARMING
A Personal Guide to Causes and Solutions

by
Sneed B. Collard III

Published by Lifelong Learning, Inc.

Printed in HK by Mantec Production Company

To my son Braden and his generation of earth stewards.

- Love, Dad

Global Warming: A Personal Guide to Causes and Solutions

Published by Lifelong Learning, Inc.
40 Second Street East, Suite 249
Kalispell, MT 59901
Phone: 877-502-7477
Fax: 406-758-6444
criss@projectcriss.com
www.projectcriss.com

Cover and Book Layout and Design:
Tim Braun/Fossil Creative
Kalispell, Montana
timbraun@me.com
www.fossilcreative.com

Cover photos: (main) A view of the Indian Ocean from space clearly shows the fragile
and finite nature of earth's atmosphere. ~ NASA Johnson Space Center
(inset) A modern electricity-generating windmill in Judith Gap, Montana. ~ Sneed B. Collard III
(insets back) Smokestack, Coalstrip Montana. ~ Sneed B. Collard III
Lightning and Thunderstorm, USA. ~ Alan Smithee
A modern electricity-generating windmill in Judith Gap, Montana. ~ Sneed B. Collard III

Library of Congress Control Number: 2010936333

ISBN: 978-0-9785367-7-0

TABLE OF CONTENTS

INTRODUCTION:
A PLANET IN HOT WATER

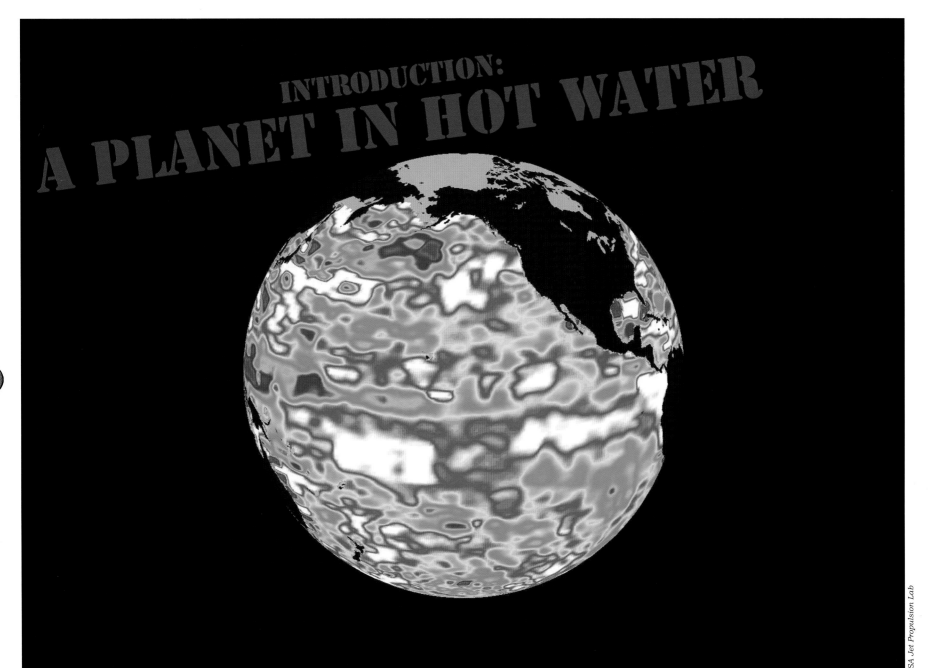

NASA Jet Propulsion Lab

Image from NASA's Jason oceanography satellite during 2002 El Niño event. White and red colors indicate warmer than normal sea surface temperatures. Blue indicate cooler than normal temperatures.

OTHER THAN WARFARE—OR MAYBE

the Super Bowl—it's hard to find a more dramatic topic than global warming. On one hand, global warming is causing worldwide calamities, including increasingly violent weather, rising sea levels, disappearing glaciers, and extinction of species. On the other hand, global warming is spurring changes in society that will lead to better ways of doing things and a safer, healthier planet. These include revolutionizing our production of energy, our transportation, and even our cities.

If we as a species are to make the changes we need to solve the problem of global warming, the first thing we need to do is understand it, and that's not easy. Global warming is a complicated subject, perhaps the most complicated you've ever run into. It involves science, economics, politics, and most of all, human behavior.

I have studied scientific and environmental issues all of my life and have written about them for almost thirty years. In all of that time, I have never encountered an issue as important and challenging as global warming. I wanted to write this book to try to sort out the many aspects of this complex issue in ways that are easier to understand. This book spends very little time trying to convince you that global warming is happening. The scientific evidence for that is overwhelming. Instead, I focus on what we need to do to solve this growing crisis. And I did use the word "we." Global warming is not something that can be fixed by scientists or politicians

Sneed B. Collard III

One consequence of global warming is that snows are melting earlier than normal. This can lead to longer summer droughts, more intense fire seasons, and warmer stream temperatures that can devastate fish populations.

or even TV personalities. It is up to each of us to make a difference. It involves the choices you and I make on how to live, travel, work, and perhaps most of all, vote. I wrote this book so you can make solid, informed choices.

Because, like it or not, that is the only way to ensure we tackle this problem before it tackles us.

GLOBAL WARMING OR CLIMATE CHANGE?

Which sounds scarier to you: global warming or climate change? The terms are often used interchangeably, but they do not mean the same thing. According to NASA and the Environmental Protection Agency, global warming is the increase in average temperatures for the entire surface of the earth and/or earth's atmosphere. Climate change, on the other hand, refers to the specific long-term changes in temperature, precipitation, sea level, and wind that occur at different places around the globe. The two terms are closely related because it is global warming that is causing the climate changes we are so worried about. In fact, increasing temperatures alone are not our primary concern. It's the many changes in weather and climate that those temperatures trigger.

Ironically, climate change sounds less scary than global warming. While global warming sounds ominous and threatening, climate change sounds, well, almost natural. This feature is not lost on environmentalists, oil companies, politicians, and even writers like myself. In this book, for instance, I usually use the term global warming because it more accurately reflects the enormous danger we all face and is the root cause of worrisome changes in climate. However, you may hear others—oil companies, for example—emphasize climate change because they do not want you to feel as worried. Not surprisingly, they also want to keep their profits intact. Knowing how the two terms are connected will help you understand and observe the underlying politics of this issue.

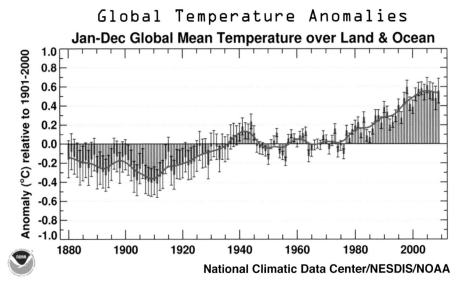

The term "temperature anomaly" means a departure from a reference value or long-term average. A positive anomaly indicates the observed temperature was warmer than the reference value, while a negative anomaly indicates the observed temperature was cooler than the reference value. The black bars show statistical standard deviations for each temperature mean.

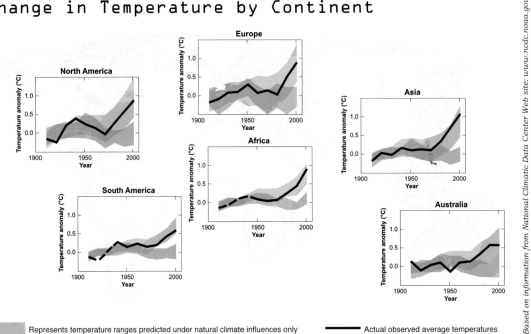

Based on information from National Climatic Data Center Web site: www.ncdc.noaa.gov

John Kohler

Visible trails of condensed water vapor, made by the exhaust of aircraft engines.

Why Not a War on Water Vapor?

Water vapor is the most important greenhouse gas, accounting for more of earth's heating than even carbon dioxide. Like other greenhouse gases, its concentration in earth's atmosphere has been rising. So why don't scientists focus on water vapor instead of carbon dioxide when they talk about global warming? Answer: because rising water vapor levels are an indirect result of increasing temperatures. This is how it works:

1) *Higher CO_2 levels trap heat in the atmosphere and make the air warmer.*
2) *Warmer air holds more water vapor and also increases evaporation rates.*
3) *Higher amounts of water vapor trap even more heat in the atmosphere, driving temperatures still higher.*

The result is that CO_2 and water vapor are part of a positive feedback loop. As CO_2 increases, so does water vapor, trapping even more heat. Scientists still have a lot to learn about this, but here is the bottom line: we can't control water vapor levels. We can only control concentrations of other greenhouse gases and hope that the water vapor problem takes care of itself.

Lynn Havens

Water vapor is both a cause and consequence of global warming. Water vapor traps more heat in the atmosphere than any other greenhouse gas. As our atmosphere warms, however, it is able to hold even more water vapor, causing temperatures to rise even further. This creates a positive feedback loop, which can only be controlled by trying to reduce other greenhouse gases.

CHAPTER ONE
GLOBAL WARMING:
THE BASICS

6

Montane forests such as this Costa Rican cloud forest are among the world's many ecosystems vulnerable to climate change.

IN LATE SUMMER OF 2007, something astonishing happened in the Arctic Ocean. For the first time in recorded human history, an ice-free passageway opened up in the waters above North America, providing a new shipping shortcut between the northern Atlantic and Pacific oceans. Canada and other interested nations immediately began bickering over who should control this valuable route. But scientists and many others saw something ominous in the new waterway—evidence that global warming may be changing our planet much more quickly than anyone expected.

THE ISSUE:
Our planet is heating up at an increasing rate. The Intergovernmental Panel on Climate Change, or IPCC, is an international body of hundreds of climate experts from around the globe. It was set up by the World Meteorological Organization and the United Nations Environment Programme. In 2007, the scientists of the IPCC released a report stating that earth's average surface temperature rose 1.33 degrees Fahrenheit (.74 degrees Celsius) between the years 1906 and 2005. That may not sound like much, but this figure is an average for the entire planet. Many regions, including huge areas of the northern hemisphere, have experienced much larger temperature increases. Temperatures in many Arctic regions, for example, have risen between 3.6 and 6.3 degrees F (2.0-3.5 degrees Celsius). Worldwide, the years 1995 through 2008, ranked as 13 of the 14 warmest years on record. What's more, the IPCC predicted that global warming in the coming century will be even more dramatic. In a few parts of the world, these warmer temperatures may actually lead to changes that make life better for people. For billions of others, global warming will not only make survival more difficult, it will damage ecosystems necessary to support life. An overwhelming number of scientists agree that humans are to blame for this unfolding crisis.

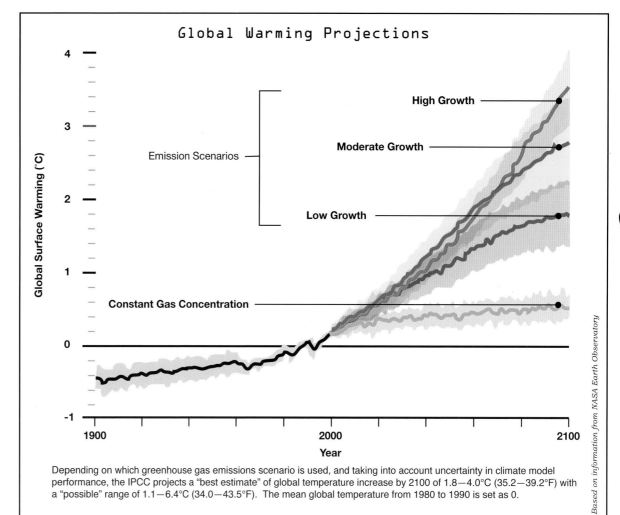

Depending on which greenhouse gas emissions scenario is used, and taking into account uncertainty in climate model performance, the IPCC projects a "best estimate" of global temperature increase by 2100 of 1.8—4.0°C (35.2—39.2°F) with a "possible" range of 1.1—6.4°C (34.0—43.5°F). The mean global temperature from 1980 to 1990 is set as 0.

Graph showing model predictions for how much earth's temperatures will rise in the coming century. The different predictions are based on how much—or how little—humans can reduce our production of greenhouse gases.

Based on information from NASA Earth Observatory

BACKGROUND FILE:

Earth's average temperatures have risen and fallen repeatedly throughout history. Studies of ice cores in Antarctica reveal that in the past 740,000 years, our planet has experienced eight major ice ages and eight warming periods. Many factors affect climate and temperatures. These include the orbit and tilt of the earth, energy output from the sun, ocean currents, and even volcanic eruptions. A critical influence on temperatures is the presence of greenhouse gases in the atmosphere.

Greenhouse gases include water vapor, carbon dioxide (CO_2), and methane. These gases trap heat in the atmosphere and make our planet a hospitable place to live. Without them, earth's average temperature would be almost sixty degrees (Fahrenheit) colder than it is today. But Earth can get too much of a good thing—which is what's happening now.

Since the start of the Industrial Revolution in the 1700s, humans have been releasing enormous amounts of CO_2 and other greenhouse gases into the atmosphere. How? Primarily by burning carbon-based fossil fuels, such as coal, oil, and natural gas. These fuels generate electricity, provide heat, and power billions of cars, trucks, and other vehicles. Farming, livestock, and destruction of forests also produce greenhouse gases.

Between the early 1700s and the year 2005, carbon dioxide levels increased about 36 percent—280 parts per million (ppm) to 380 ppm—the highest by far in the last 650,000 years. As our appetite for fuel and energy grows, we are dumping more and more greenhouse gases into our atmosphere every year. The higher temperatures these gases create are already unleashing a series of calamities across the planet.

According to the Center for Global Development, the world's 50,000 power plants dump about 10 billion tons of carbon dioxide into the atmosphere every year.

Sneed B. Collard III

A BRIEF, HOT HISTORY OF GLOBAL WARMING

Jean Baptiste Joseph Fourier not only has a very long name, he can be considered the father of global warming. Back in 1824, this French mathematician discovered that Earth's temperature was slowly increasing. He blamed this increase on heat trapped in our atmosphere. It wasn't until the 1950s, however, that scientists began seriously measuring levels of greenhouse gases and exploring whether or not humans were responsible for increasing temperatures.

Throughout the 1980s, scientific evidence for global warming was building, along with serious concern that we needed to do something about it. By 1988, the countries of the United Nations had become worried enough to assemble a group of scientists to evaluate the problem. That group was the Intergovernmental Panel on Climate Change (IPCC). In 1992, 150 nations—including the United States—signed a United Nations declaration committing themselves to reducing greenhouse gases. In 1997, leaders from around the world drafted the Kyoto Protocol treaty—a legally-binding agreement to lower greenhouse gas production. By 2005, more than 100 nations approved, or ratified, the document. The United States—until recently, the world's largest emitter of greenhouse gases—was not one of them. Why not?

The Kyoto Protocol allowed China, India, and other developing nations to release unlimited amounts of greenhouse gases. The United States stated this provision was unfair and used this as a reason for not signing the treaty. Developing nations, on the other hand, correctly stated that it was primarily the United States and other industrialized nations that had created global warming. Why, they argued, should poor nations be punished for wanting to catch up and improve their economies?

On the face of it, both sides had reasonable arguments. What wasn't reasonable is that the United States used Kyoto as an excuse for taking almost no action on the problem. While European nations aggressively worked to cut their greenhouse emissions, the United States carried on business as usual. In fact, between 1990 and 2005, greenhouse emissions in the United States actually increased by 16.3 percent.

Fortunately, America's leaders are beginning to take action. In 2009, President Obama signed the American Recovery and Reinvestment Act, a bill loaded with money and tax incentives to promote clean energy and conservation investment. Unfortunately, as this book goes to press, Congress has still failed to pass long-term, comprehensive legislation to set clean energy standards and reduce global warming emissions. Chances for such legislation remain highly uncertain.

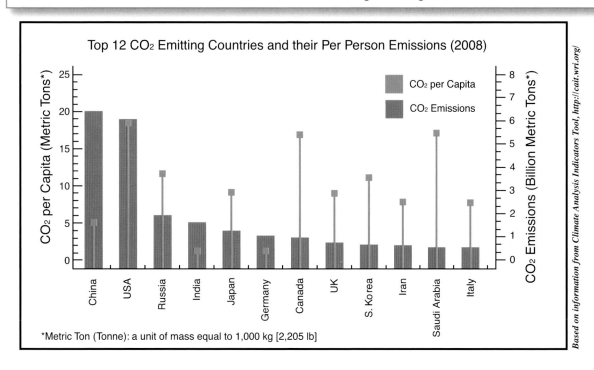

Top 12 CO₂ Emitting Countries and their Per Person Emissions (2008)

*Metric Ton (Tonne): a unit of mass equal to 1,000 kg [2,205 lb]

Based on information from Climate Analysis Indicators Tool, http://cait.wri.org/

DATA FILES:
Global Warming's Impacts

Global warming's negative impacts promise to overshadow almost every aspect of our lives. These impacts include:

- Rising sea levels.
- Increased number of extreme weather events (drought, flooding, storm damage).
- Lower food production.
- Extinction of species.
- Destruction of forests.
- Increased spread of harmful, or invasive, species.
- More human health issues.

A look at a few of these begins to paint the picture.

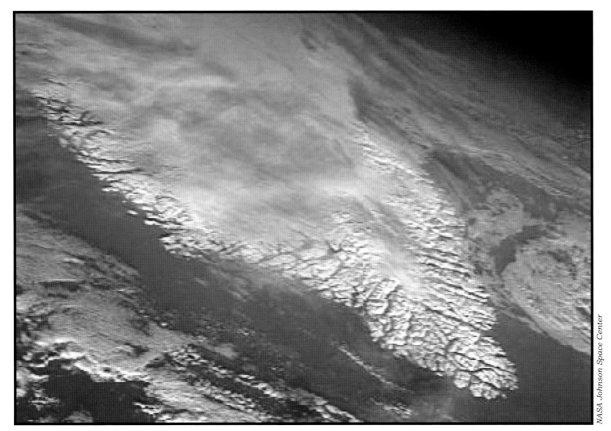

(Above) The Greenland ice sheet alone holds enough water to flood huge portions of our planet.
(Lower Right) Global warming data suggest an increase in extreme weather events such as Hurricane Katrina, shown in this image created from satellite data.

Rising Sea Levels

On December 24th, 2006, *The Independent* news service reported, "Rising seas, caused by global warming, have for the first time washed an inhabited island off the face of the earth." Lohachara Island was located in the vast Ganges river delta between Bangladesh and India, and was home to 10,000 people. In 2006, it vanished under rising sea levels. Yet, Lohachara was only the first of hundreds of islands worldwide that face obliteration. Since 1993, the global sea level has risen about 3.1 millimeters per year—or about one and a quarter inches per decade. The rate is increasing.

Two processes cause sea levels to rise. As water gets warmer, it expands, taking up more space. Melting glaciers and other "permanent" ice dump even more water into the oceans, pushing up sea levels further. The Greenland ice sheet alone holds enough water to raise global sea levels by 23 feet (7 meters). Scientists do not expect the entire ice sheet to melt anytime soon, but already, Greenland is losing 100 billion tons of ice each year.

According to one scientist studying the problem, by the year 2100, Greenland's melting ice could push up sea levels by as much as three feet (about one meter)—inundating low-lying coastal areas around the world, from much of Florida and California to heavily populated river deltas in Asia, Africa, and the Americas. On Pacific Island nations, such as Vanuatu and Kiribati, entire villages have already been forced to move to higher ground. If sea levels rise as predicted, millions of people worldwide will have to relocate as their homes disappear beneath the waves.

Extreme Weather Events

According to the Intergovernmental Panel on Climate Change, it is "virtually certain" that earth's land areas will experience fewer cold days and nights and more frequent hot days and nights

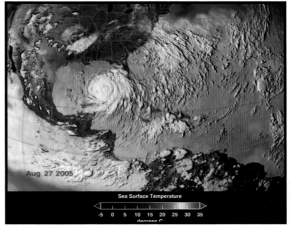

in the future. Scientists predict heat waves and torrential rains are "very likely" to increase and that it is "likely" the number of droughts and extremely intense hurricanes will increase. These processes have already begun. In many mountainous areas, winter snow is melting earlier and more quickly than it used to, leading to flooding rivers and reduced water supplies during summer's hottest months. Food production in many parts of the world is falling because of drought, storm damage, and soil erosion. Even human health is suffering. In China, authorities believe hotter temperatures may be causing as many as one million additional deaths per year through strokes and heart disease. Mosquitoes are also thriving under warmer temperatures, adding to the range and incidence of malaria, dengue fever, and West Nile virus in Africa, Europe, Asia, and elsewhere.

Extinction of Species

On May 14th, 2008, the U.S. government officially listed the polar bear as an endangered species. The reason? Polar bears need summer sea ice to hunt and rest on, but sea ice cover has shrunk dramatically over the last three decades. Without the ice, polar bears are already starving and drowning in record numbers. Yet the polar bear is only one of millions of species facing extinction on our warming planet. As the planet heats up, many tropical rainforests and other forests will become grasslands, while grasslands turn into deserts.

Just how many species we will lose remains uncertain. Some scientists believe we may lose a relatively small number. IPCC scientists, on the other hand estimate that up to 30 percent of all species risk extinction—a loss that would forever change the face of our planet.

Warmer arctic temperatures have already impacted the polar bear, which needs summer ice to hunt seals.

NOAA

CONCLUSION:

The solution to global warming is simple: we must dramatically reduce the amount of greenhouse gases we release into the atmosphere. There's only one way to do that: quit burning coal, oil, and natural gas for energy production. Unfortunately, that is not as easy as it sounds. The rest of this book will be devoted to examining the details of what's involved.

Coral Reefs Taking a Bath

Coral reefs have shown some of the earliest—and most harmful—effects of global warming. Extremely warm temperatures lead to episodes of coral bleaching in which corals dump the tiny algae called zooxanthellae living inside of them. These zooxanthellae produce much of the food that a coral needs to survive. If coral bleaching is mild, some corals can recover from bleaching, but often they die. As the ocean heats up, however, coral bleaching events around the world have been more severe and more frequent. Six major coral bleaching events have occurred in the past twenty years. In the 1997-98 event, 46 percent of the corals were killed on some Indian Ocean reefs. In 2001-02, large areas of Australia's Great Barrier Reef were also affected. Scientists believe that, unless global warming is rapidly reversed, rising ocean temperatures will destroy most of the world's coral reefs before the year 2050. As the reefs disappear, they will take thousands of other species with them—and deprive millions of fishermen and other coastal people of their livelihoods.

Sneed B. Collard III

CHAPTER TWO
ENERGY:
THE KEY TO GLOBAL WARMING

Nations around the world are turning to wind farms as a cheap, clean source of energy.

Sneed B. Collard III

IN MAY 2008, LEGENDARY OIL TYCOON T. Boone Pickens announced his latest plan to make money from the booming energy industry. One of the wealthiest people in America, Pickens had made his $3 billion fortune from oil and natural gas. Surprisingly, his newest venture did not involve oil, gas, or any other fossil fuel. Instead, he planned to build a gigantic wind farm on the plains of west Texas. Pickens later scrapped his plans, but his original announcement did more than shock observers. It provided a real-life metaphor for how the world is scrambling to deal with the dual crisis of energy and global warming.

THE ISSUE:

The pursuit and production of energy for electricity, heating, and transportation is the leading cause of global warming. In the United States, 99 percent of the carbon dioxide our society produces comes from energy production. How quickly we transform our energy and transportation systems may be the most important factor in determining whether or not our planet—and our civilization—survive.

BACKGROUND FILE:

Fossil fuels powered the industrial revolution and enabled humans to create the modern societies we see today. Hydroelectric power, nuclear power, and even burning wood have made their contributions, but it is coal, oil, and natural gas that have truly transformed society. Today, these three fossil fuels account for 80 percent of the energy we produce and consume.

Coal was the first important fossil fuel. By the Middle Ages, it was widely used in Europe for heating and small-scale industrial processes. By the nineteenth century, it was used to power steam engines and produce steel. Today, more than 90 percent of the coal that we consume is burned to generate electricity. Because it is so abundant and cheap, it is the most common source of energy on the planet.

Since the end of World War I, however, oil (petroleum) and natural gas have played increasing roles in energy production. Oil, especially, has become the dominant player in transportation because it is used to make the gasoline that powers the world's 700 million cars. It also powers most airplanes, ships, and other vehicles. In fact, oil and transportation are so closely linked that you cannot discuss one without the other. Natural gas, on the other hand, is more

Sneed B. Collard III

Every day, dozens of mile-long coal trains cross the United States in an effort to keep up with our enormous demands for electricity. The locomotives pulling these trains are themselves a major source of U.S. CO_2 emissions.

associated with heating, cooking, and electricity generation.

The use of all three fossil fuels has raised living standards for many people on the planet. It has brought us the electricity that runs our appliances, computers, industrial machines, and lights. It has provided the fuel we need to heat our homes and drive our automobiles and fly to Hawai'i for vacations. It has made industrial processes cheaper and more efficient, producing an abundant supply of clothes, electronics, and other goods that we want. Last, and perhaps most important, it has powered the vast agricultural operations that feed most of earth's population.

We need energy for so many things that our demand for fossil fuels has skyrocketed. In the United States, energy consumption in the year 2000 was nineteen times what it was in 1949. But our insatiable energy appetite has placed us on the brink of global disaster by raising earth's temperatures. It has also created other massive problems on our planet.

DATA FILES:
Fossil Fuel Problems

Global warming is not the only drawback to burning fossil fuels for energy. Like a spider web, fossil fuel problems penetrate almost every aspect of modern society. Here's but a sample . . .

Ecosystem Damage

Oil- and coal-related activities have seriously damaged almost every region of the globe. Until recently, America's most famous fossil fuels disaster was the *Exxon Valdez* oil spill, which dumped 10.8 million gallons (40.8 million liters) of crude oil into Alaska's Prince Williams Sound. The spill killed approximately a quarter of a million seabirds along with thousands of marine mammals. Toxic effects from the spill persist today.

On April 20th, 2010, however, the British Petroleum drilling platform *Deepwater Horizon* exploded and sank in the Gulf of Mexico. The accident killed eleven workers and triggered the worst oil spill in history. By August, the United States government estimated

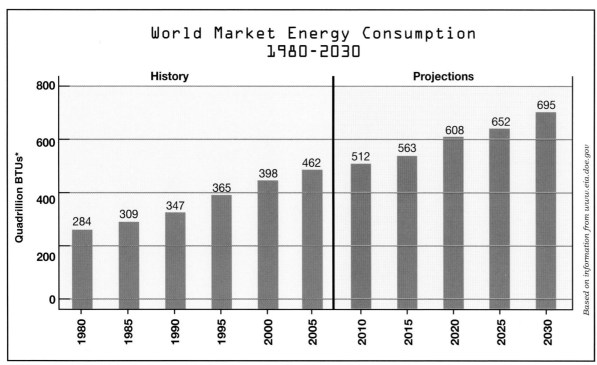

*BTU stands for British thermal unit—a standard measure for energy.

Despite heroic clean-up efforts, oil spills often have catastrophic, long-term impacts on ecosystems.

that 4.9 *million* barrels of oil had been spilled—more than 200 million gallons (780 million liters). At its height, the oil slick covered about 29,000 square miles (75,000 square kilometers) of ocean surface, closing more than one-third of Federal fishing grounds in the Gulf. From Louisiana to Florida, toxic crude oil battered shorelines, including many sensitive wetlands areas. By September 15th, the U.S. Fish & Wildlife Service had counted 5,939 dead seabirds, 584 dead sea turtles, and 92 dead marine mammals along the coast, many of which showed visible signs of oil. Total cost of oil cleanup and compensation to victims of the spill was estimated to exceed $20 billion, but the damage to the Gulf of Mexico economy and environment promised to last for decades.

As shocking as these statistics are, ongoing oil-related activities have been even more destructive. In 1971, in the Amazon Basin of Peru and Ecuador, the American oil company Occidental Petroleum began building oil installations across more than a million acres of pristine rainforest. The company took few precautions to protect the environment. Oil drilling not only led to numerous oil spills, over a thirty-year period it produced nine billion gallons (34 billion liters) of wastes loaded with lead and other toxic chemicals. These wastes were dumped directly into rainforest streams and rivers. Thousands of acres of forest were also cleared for 300 miles (483 kilometers) of roadways and a 532-mile (856-kilometer) pipeline to transport the oil. The Peruvian government

Louis Helbig   www.beautifuldestruction.ca

Some environmental groups have labeled Alberta's tar sands activities the most destructive project on earth. Part of what makes the project so harmful is that it takes tremendous amounts of energy to extract oil from the sands.

acknowledged that oil activities had reduced the region to the most environmentally damaged part of the country. However, that has not stopped the government from recently opening more than 100 million additional acres (40.5 million hectares) of unspoiled rainforest to oil exploration.

Closer to home, in Alberta, Canada, the rush is on to develop oil-soaked "tar sands" by stripping away sensitive boreal forest. To reach the tar sands, companies have to excavate huge areas of land, completely obliterating the forest ecosystem. Already, about 50,000 acres (about 20,000 hectares) have been destroyed. Plans call for another 740,000 acres (about 300,000

hectares) to be "developed." An even greater problem is that processing the tar sands devours enormous quantities of Alberta's fresh water and leaves behind billions of gallons of polluted, toxic wastes that are leaking into the surrounding environment.

Pollution

Energy-related activities are responsible for almost all man-made air pollution. In the United States, personal vehicles spew out more than half of the deadly carbon monoxide emissions that we produce. Cars and other vehicles also produce 56 percent of the nitrous oxides and 45 percent of the volatile organic compounds that produce ground-level ozone—an unhealthy chemical that harms humans and damages plant life. A nonprofit group called the Clean Air Task Force estimates that small particles released from diesel engines lead to 21,000 premature deaths each year in the United States.

According to a 1995 report compiled by the Smithsonian Institution, almost 700 million gallons (2.6 billion liters) of oil are dumped into the world's oceans annually. This is the equivalent of about three-and-a-half *Deepwater Horizon* disasters every single year. Only about 5 percent of the total comes from major oil spills, however. More than half—363 million gallons—comes from oil dumped down drains from car oil changes and from spilled oil washed into storm drains and sewers from roadways, parking lots, and other land areas.

Global Conflict

Across the globe, exploration and control of oil supplies has generated enormous, often violent, conflict. Many experts—including former Secretary of State Henry Kissinger and former Federal Reserve Chairman Alan Greenspan—believe the Iraq War was largely a fight over that country's huge oil reserves. Other, smaller conflicts occur throughout the world as local people battle huge energy corporations for control of their own lands and lives.

To safeguard our oil supplies, Western nations support a number of dictatorships with terrible records of human rights abuses. These include Nigeria, Myanmar, and Saudi Arabia. Even worse, the money we spend on oil is often funneled to terrorist groups. Al Qaeda—the group responsible for the September 11, 2001, attacks—received much of its support from wealthy Saudi Arabians. Where did these Saudis get their money? You guessed it—from oil sales. Our thirst for oil has put our nation and the world in a high-stakes balancing act, one that is proving more and more costly and unstable.

CONCLUSION:

Because of the number and magnitude of problems associated with fossil fuels, we must start replacing them with other energy sources. Doing so will help reduce global warming, but will also help solve a number of other environmental and political problems.

In fits and starts, that process is slowly shifting out of first gear.

Jesse B. Awalt, U.S. Navy

A desire to control energy supplies has contributed to numerous armed conflicts around the world, at huge personal and economic cost.

PEAK OIL

In the 1950s, a famous geologist, M. King Hubbert, recognized that, at some point, oil production would peak and then decline. In the United States, he predicted, the peak would occur between 1965 and 1970. Domestic oil production, in fact, did begin declining in 1970 and 1971. Later, Hubbert predicted that global oil production would peak between 1995 and 2000. It appears the peak occurred in 2005.

So what?

Peak oil signifies the point at which global demand for oil exceeds the global supply. Many experts believe this event may trigger a whole series of negative effects. According to a 2005 report commissioned by the U.S. Department of Energy, "as peaking is approached, liquid fuel prices and price volatility will increase dramatically and, without timely mitigation, the economic, social, and political costs will be unprecedented." In other words, the world economy and society could well be thrown into chaos.

We may already be seeing the beginnings of peak oil's impacts. In recent years, the demand for oil in China and India has grown rapidly. With more nations competing to buy oil, the price of oil has soared. Between September 2003 and June 2008, oil increased from about $25 per barrel to more than $135 per barrel. For the first time in history, many Americans began paying more than $4 per gallon for gasoline.

Since then, oil's price has fluctuated wildly. These price changes ripple through every aspect of society from the cost of food to the price of medical treatment, lawn care, building materials, and family vacations. Is American or world society on the verge of chaos? Probably not. After all, Europeans already pay seven to eight dollars a gallon for gas. Much of this cost comes from government taxes that have discouraged larger cars—and led Europeans to be more energy efficient.

The problem is that Americans have not had to be energy efficient. How would we cope if our gas also suddenly shot up to six or eight dollars a gallon?

It's something we should think about—and plan for.

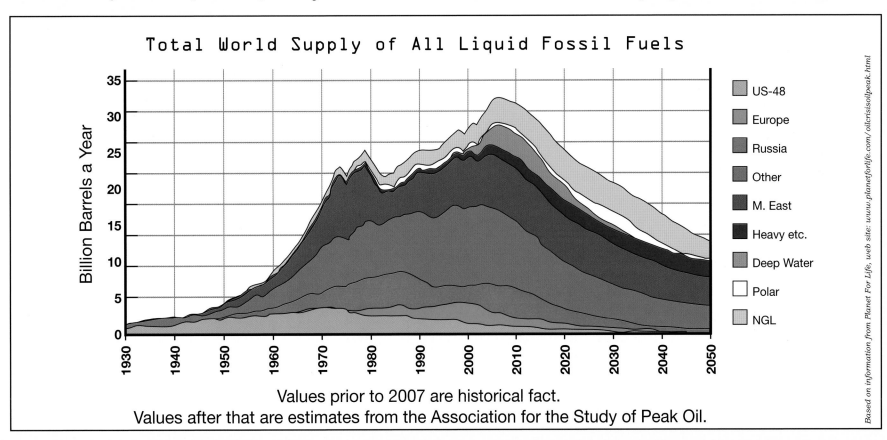

Total World Supply of All Liquid Fossil Fuels

Billion Barrels a Year

Legend: US-48, Europe, Russia, Other, M. East, Heavy etc., Deep Water, Polar, NGL

Values prior to 2007 are historical fact.
Values after that are estimates from the Association for the Study of Peak Oil.

Based on information from Planet For Life, web site: www.planetforlife.com/oilcrisisisoilpeak.html

18

Sneed B. Collard III

Government subsidies have helped expand the production of ethanol from corn, but not without serious environmental and humanitarian concerns.

AT THE END OF SEPTEMBER 2008, the ocean became just a little more valuable. That's when the world's first commercial "wave farm" began generating electricity. The new electricity plant resembles three enormous red snakes floating three miles (five kilometers) off the coast of northern Portugal. As passing waves lift each giant tube up and down, they drive pistons and motors inside. These generate electricity that is fed to shore through a cable on the sea floor (see page 22). Together, the three giant tubes generate 2.25 megawatts of electricity—enough to power about 1500 homes. Plans are underway to expand the facility and build additional wave generation plants worldwide.

THE ISSUE:

As the need to replace fossil fuels becomes urgent, many energy alternatives are being considered. Some, however, have high hidden costs for our environment, global security, and even our pocketbooks. Which new energy sources we develop will largely determine how successful we are at combating global warming and developing a healthier, safer economy.

BACKGROUND FILE:

Many replacements for fossil fuels are being proposed or developed around the world. Some promise to create as many problems as they solve. Ethanol is a good example. Ethanol is an alcohol that can be used as fuel for motor vehicles. Although ethanol can be made from sugar cane and other plant materials, most U.S. ethanol is made by fermenting corn.

Politicians, farmers, and large agriculture and chemical corporations have pushed ethanol as a partial replacement for gasoline. In the United States, however, ethanol takes almost as much—if not more—energy to make as it yields when burned. Because ethanol must be "grown," it also robs us of valuable farmland just at a time when the world is being hit with major food shortages. According to Corn Growers Association, by the year 2015, one-third of the U.S. corn crop will be used to produce ethanol. Already, this has made corn more scarce and pushed up food prices. The USDA downplays ethanol's impact on food prices, but a 2007 study released by Iowa State University revealed that because of ethanol, Americans were already paying $14 billion more for food annually.

Other people and organizations—especially from coal mining states—push burning more coal to produce electricity. Coal, though, has a long list of drawbacks for environmental safety and human health (see sidebar: IS THERE SUCH A THING AS CLEAN COAL?). Hydrogen and nuclear power are also being advocated by particular groups, but hydrogen faces major technological hurdles to become useful, while nuclear power comes with huge environmental, safety, and security costs (see the debate on pages 24 and 25).

Fortunately, we do have safe, affordable choices that protect the environment and are already creating thousands of new jobs for people.

Sneed B. Collard III

Even some gas station owners recognize the drawbacks of planting corn for fuel. Here, an owner in Missoula, Montana, plants his own corn crop in protest.

A LITTLE ABOUT HYDROGEN

The promise of hydrogen is not actually as a fuel, but as an energy carrier. As such, it is stored in units called fuel cells. Inside the fuel cell, hydrogen gas is separated into protons and electrons. This generates an electric current that, like a conventional battery, can power an electric motor. Before fuel cells can begin to power automobiles, however, the costs to produce hydrogen and fuel cells must be lowered dramatically. A large—and expensive—grid of hydrogen filling stations must also be created. According to a 2004 analysis by the Union of Concerned Scientists, hydrogen fuel cell vehicles will not be ready for widespread use until about 2025. Even if that time can be cut in half, it may be too long—and expensive—to head off the worst of global warming's impacts.

IS THERE SUCH A THING AS CLEAN COAL?

During the past several years, Americans have been seeing more and more advertisements about the benefits of "clean coal." These advertisements not only state that coal is cleaner than it used to be, but also suggest that burning coal will not release greenhouse gases into the atmosphere. Not surprisingly, these advertisements are paid for by a coalition of corporations heavily involved in the coal industry. All of these corporations will earn handsome profits if the world keeps burning coal to produce electricity. The question is, are claims of "clean coal" true?

Even if we ignore coal-related disasters such as coal sludge spills that have devastated river systems, a little research reveals that the answer to this question is basically "no."

Burning coal releases some of the most toxic forms of air pollution. These include sulfur dioxide, which produces the acid rain that has poisoned forests, farmland, lakes, and rivers around the world. It includes nitrogen oxides, which contribute to acid rain and also pose a significant health hazard to people in the form of smog. Coal-fired power plants are responsible for emitting most of the mercury that is poisoning ocean food webs. Pollution from coal-fired power plants is so bad that the Environmental Protection Agency ordered the industry to install pollution control devices to reduce harmful emissions. These efforts have been worthwhile. Today, coal-fired plants still release large amounts of air pollution, but not as much as they used to. So in one sense, there is "cleaner coal," but not "clean coal."

The coal industry's promises to eliminate greenhouse gases are even more misleading. It's true that industry scientists are working on ways to capture carbon dioxide from coal-fired power plants. One idea calls for pumping and storing this carbon dioxide underground. As I write this, however, there is not a single commercial coal-fired power plant anywhere in the world that captures all of its carbon dioxide and prevents it from escaping into the atmosphere. What's more, even if it does prove possible to capture and store carbon

dioxide, the industry is decades away from being able to do it on a large scale.

A little investigation reveals that the main benefit of the Clean Coal campaign is not as a solution to global warming. It's to demonstrate how effectively advertising can confuse people about the truth behind one of the world's dirtiest industries. When evaluating any claims of clean energy, here is a good rule of thumb to follow: If someone has to burn something to create energy, that source of energy <u>will</u> create greenhouse gases and is <u>not</u> clean.

NOISE TO SIGNAL
RobCottingham.ca

Yes, but it's <u>clean</u> coal.

DATA FILES:
Wind, Solar, and Wave Power

Wind Power

Wind power is the first new clean, safe, renewable energy to have a major impact on the world's energy supply. Thanks to government subsidies, unpredictable oil prices, environmental concerns, and an abundance of wind on the planet, wind farms are sprouting like forests across the globe. Germany has long been a leader in wind energy. Today, it produces about 7 percent of its electricity from wind farms. In 2008, however, the United States surpassed Germany and now produces about one-fifth of the world's wind energy. China has also surpassed Germany, and in 2009, doubled its wind energy production for the fourth year in a row. Analysts believe that by the year 2025, wind will produce about 12 percent of the world's energy needs.

Photovoltaic, or PV, systems can be installed almost anywhere the sun is shining. As costs come down, we will see more rooftop PV units like this one.

Solar Power

Like wind, solar power is radiating across the planet. There are two distinct types of solar power. One is photovoltaic solar power (PV), which uses silicon to directly convert solar rays into electricity. The second is solar thermal power, which relies on the sun to heat up water or other liquids. These hot liquids can be used directly for home and industrial needs or can be used to produce electricity by powering steam turbines.

Photovoltaic, or PV, panels are still relatively expensive to produce, but world production quadrupled between 2005 and 2009. Solar thermal installations are also increasing because they are cheap and easy to install. Germany, China, and Taiwan are all major players in solar energy. The PV industry employs 40,000 people in Germany and more than 246,000 in China. In 2009, China and Taiwan together produced more than 5,000 Megawatts of new PV—more than the entire world produced in 2007. So far, China exports most of its PV panels to other countries. But China already has more than 40 million solar thermal systems in place in its own country.

The advantages to both wind and solar energy over fossil fuels are irresistible:
1) There is practically a limitless amount of both wind and sunshine on the planet.
2) These energy sources are renewable. They will never run out.
3) Once they are manufactured and installed, wind and solar systems produce NO carbon dioxide or other pollutants.

Giant wave generators, like these off the coast of Portugal, may soon become a common sight in the world's coastal regions

Not that wind and solar power are problem-free. When it comes to wind farms, for instance, NIMBY ("Not In My Back Yard") voices often turn shrill. Ted Kennedy, one of America's most liberal politicians, repeatedly fought the construction of a wind farm on Horseshoe Shoal in Nantucket Sound near the Kennedy compound on Cape Cod. More objective critics jump on the fact that the wind doesn't always blow and the sun doesn't always shine. While true, this simple-minded argument overlooks the fact that with a diverse grid of renewable energy sources, power will almost always be available somewhere. Also, no one is saying that wind and solar power have to supply all of our energy needs. Natural gas, coal, hydroelectric, and other power sources can always make up any shortfall we experience.

Wave Power

Though still in its early stages, wave power also has potential to provide us with clean energy. Worldwide, the power contained in waves is enormous. The problem is that the wave energy is widely distributed. In other words, one needs large machines to intercept and capture wave energy. That has not stopped people from pursuing it. Portugal plans to install more of the giant red "snakes" in the near future. This will bring its total wave energy production to 21 Megawatts.

What About Hydroelectric Power?

In many ways, hydroelectric power was the first large-scale renewable energy, powering much of America's early industries. Hydroelectric power is generated by water rushing through turbines that generate electricity. This usually requires building dams to create a consistent supply of water. Today, "hydro" generates about 19 percent of the world's electricity. It emits no greenhouse gases directly, but producing the energy to build the dams does release large amounts of CO_2. Dams also have other serious drawbacks. One is that new dams and hydro generating stations are extremely expensive to build. Another is that, at least in the United States, we are running out of locations to build new hydroelectric facilities.

A third problem is that dams destroy important river ecosystems and displace people already living in these places. China's giant Three Gorges Dam cost about $24 billion to build and is the world's largest hydroelectric facility. The dam, however, required moving more than one million people during construction and has severely impacted the Yangtze River's aquatic life. Water seeping into soils around the dam has triggered hundreds of landslides that have killed farmers and drowned fisherman. The consequences of a large earthquake that could rupture the dam are horrifying at best. Although China continues with plans to build more giant dam projects, these and other problems of dam construction will severely limit the expansion of hydroelectric power in most regions around the globe.

Lynn Havens

CONCLUSION:

Taken together, renewables have vast potential to protect our planet and stimulate the economy. The German government estimates that its renewable energy sources already prevent about 100 million tons of CO_2 from being released into the atmosphere each year—the equivalent of removing about 18 million U.S. cars from the road. The Global Wind Energy Council estimates that worldwide, wind power alone already provides jobs for 400,000 people. This shows that the benefits of renewables extend beyond energy.

But expanding renewables is only one part of the global warming challenge.

Fictional Debate

Nuclear Power—Savior or Satan?

The following fictional debate is part of an ongoing series sponsored by the fictional Alliance for a Carbon-Free America. Today's topic: Can nuclear power solve our energy needs while averting catastrophic global warming? With us today are Nat Breakwater from the Electricity Consumption Institute and Shirley Wineglass of the Society for a Peaceful Planet. Our moderator tonight is Russell Grayscale from National Interdependent Radio.

Moderator: *Mr. Breakwater, many people nowadays have asserted that in a warming world, our best option is to "go nuclear" by building more nuclear power plants. Do you agree?*

Mr. Breakwater: *Good evening, and by the way, I love that tie you're wearing.*

Moderator *(blushing): Oh, uh, thank you. My daughter picked it out.*

Mr. Breakwater *(chuckling): I'll have to ask her to do my shopping.*

(Everyone but Ms. Wineglass laughs.)

Mr. Breakwater: *But in answer to your question, in the face of global warming, nuclear power is not only our best option, it is our only option. It is cheap, safe, and produces absolutely no greenhouse gases.*

Ms. Wineglass: *That absolutely is not true.*

Moderator: *We'll get to your response in a moment, Ms. Wineglass, but first Mr. Breakwater, can you explain your rather startling statement, starting with cost?*

Mr. Breakwater: *I'd be delighted. Nuclear power is among our most cost-effective energy sources, coming in at 3-4 cents per kilowatt. That's less than coal, natural gas, wind, or solar power. Furthermore, uranium, the raw fuel for a nuclear power plant, is dirt cheap.*

Moderator: *Ms. Wineglass, I can see you are itching to reply.*

Ms. Wineglass: *First of all, your cost estimates are extremely conservative. The nuclear power industry has a notorious reputation for cost overruns, averaging one billion dollars per reactor. Also, nuclear power has received more than $100 billion in taxpayer subsidies. When you figure in its true costs, nuclear power costs 9-10 cents per kilowatt compared to 2-3 cents for energy conservation measures and 6-7 cents for wind power. And that's not even figuring in the price of nuclear catastrophes such as Three Mile Island and Chernobyl.*

Moderator: *And what about safety concerns, Mr. Breakwater? All of us remember the devastation caused by the meltdown of nuclear reactors in Russia and Pennsylvania.*

Mr. Breakwater: *I'm glad you brought that up. Casualty figures from Chernobyl were grossly exaggerated by liberal opponents of nuclear power. I also want to say that Chernobyl was an unsafe Soviet reactor. Our new generation of reactors promises to be totally safe. Even experts with the Union of Concerned Scientists acknowledge that the built-in safety of new designs makes a meltdown highly unlikely.*

Ms. Wineglass: *UCS has endorsed only one of several new nuclear designs. Also,* possible meltdowns are not the only things that make nuclear power dangerous. Every part of the nuclear process, from mining and refining uranium to actual power generation, releases cancer-causing radioactive materials into the environment. Nuclear reactors in this country have left behind more than 52,000 tons of highly radioactive spent fuel that will remain dangerous for thousands of years. That's not to mention over 90 million gallons of radioactive liquid and thousands of tons of dangerous tailings from mining activities. After fifty years of debate, we still do not have safe, proper means of disposing of this deadly material.

Mr. Breakwater (snorting): Even if your figures are correct, which I doubt, this is a small price to pay to avert global warming and safeguard our energy security.

Ms. Wineglass (growing more heated): Well, speaking of security, how well are nuclear power plants protected from terrorist threats? For that matter, how well are nuclear waste stockpiles protected. All it would take is one "dirty bomb" loaded with nuclear waste to contaminate an entire American city!

Moderator: Now, let's just take a deep breath here, everyone. Mr. Breakwater, how would you care to respond?

Mr. Breakwater: While it's true that more needs to be done to protect our nuclear facilities from terrorist threats, these threats are nothing compared to the looming threat of global warming. One thing my esteemed colleague cannot argue is the fact that nuclear power releases no greenhouse gases into our atmosphere.

Ms. Wineglass: Again, not true. Producing the energy required to build a nuclear plant, mine and process the uranium, and transport it releases huge amounts of carbon dioxide.

Mr. Breakwater: Yes, but so does the manufacture of windmills and solar panels.

Ms. Wineglass: True, but windmills and solar panels do not contaminate our environment for thousands of years. They don't provide ripe targets for terrorists, either.

Moderator (adopting his serious reporter face): Well, unfortunately, our time is up. I can see we're not going to solve this debate anytime soon. But I want to thank you both for being here and for a lively discussion. And remember to tune in to National Interdependent Radio, where we need you, and we hope you need us.

Sherwin McGehee

If nuclear power proponents have their way, the giant cooling towers of new nuclear power plants may soon be rising across America's landscape. The question: is this something we should be happy about?

CHAPTER FOUR
FOCUS ON TRANSPORTATION

Amy Ratzlaf

Portland, Oregon, has for decades worked to become one of America's greenest cities.

ATLANTA, GEORGIA, IS ONE OF THE most car-dependent cities in the world. Even by American standards, Atlantans burn more fuel and spend more time stuck in traffic than almost anyone else. They have to. Only two rail lines serve city commuters. The city is so spread out that riding somewhere by bus can take hours. In July 2006, however, the Atlanta City Council set in motion a plan to begin taking the city back from the automobile. Called the "BeltLine," the plan will create a 22-mile loop around Atlanta. The loop will be filled with parks, bike trails, greenspace, and new development. Perhaps most important, a new light rail mass transit system will follow the loop, linking many of Atlanta's sprawling neighborhoods for the first time. This will allow more people to get out of their cars—and make a small, but important contribution to fighting global warming.

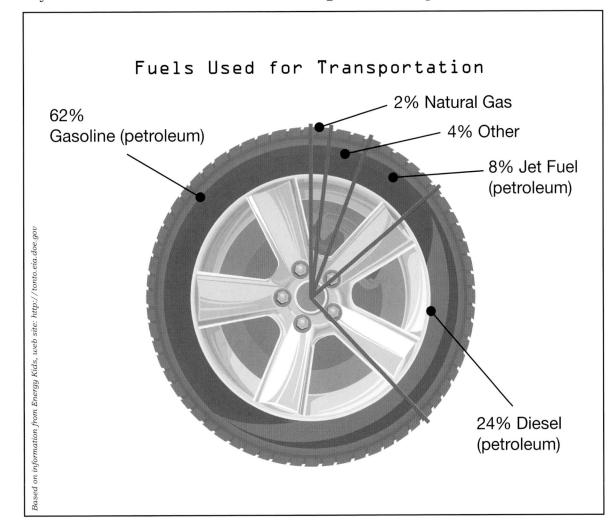

Fuels Used for Transportation

62% Gasoline (petroleum)

2% Natural Gas

4% Other

8% Jet Fuel (petroleum)

24% Diesel (petroleum)

Based on information from Energy Kids, web site: http://tonto.eia.doe.gov

THE ISSUE:
Transportation—especially cars and trucks—consumes a huge portion of the fossil fuels that are burned each year. Worldwide, one-quarter of the oil produced is burned up in gasoline engines. In the United States, the percentage is even higher. Almost half of the oil we consume goes toward our gasoline-powered motor vehicles. That's about ten million barrels—420 million gallons (1.59 billion liters)—Americans burn each and every day just to drive cars and light trucks. When you factor in airplanes, heavy trucks, and other vehicles, transportation accounts for about 68 percent of the oil Americans consume. Any effort to combat global warming will require overhauling our transportation system to eliminate gas-powered vehicles and provide better methods of transportation.

BACKGROUND FILE:
From its beginning, motorized transportation has depended on fossil fuels. The first successful steam locomotives burned coal in the early 1800s. The heat from the coal boiled water, which in turn produced the steam that pushed a piston that drove the wheels forward. Steam-powered automobiles were also tried, but these proved too heavy and complicated to be practical.

Trains dominated long-distance transportation for more than a century. Many were steam-powered, but electric trains also became popular, especially in cities. By 1900, electric subways and streetcars had begun sprouting up in many U.S. cities. In these cities, they would remain the most commonly used transport method until after World War II. During this same time, however, a new, less-efficient form of transportation was picking up speed.

Karl Benz patented the first practical automobile in Germany in 1886. By 1923, Henry Ford was making almost two million Model-T cars a year here in America. These cars were powered by gasoline made from petroleum, and they were a big hit. People loved the freedom of having their own cars. Powerful industrial interests also recognized that the automobile could make them vast profits. These interests worked hard to make cars not only convenient, but essential to American life.

Beginning in the 1930s, General Motors, Standard Oil, Firestone Tire, and other auto interests began buying up and shutting down electric trolley systems across the country. Federal and State governments also began pouring in billions of dollars to pave roads and build new highways. By the 1960s, Americans could no longer choose between public transportation and the automobile. We had to have a car if we wanted to get somewhere in a reasonable amount of time.

Today, there are more than 700 million cars in the world and another 200-plus million trucks and other motor vehicles. The United States has the largest fleet, topping 250 million motor vehicles. Almost all of these vehicles burn fossil fuels. Worldwide, motor vehicles account for about 10 percent of the carbon dioxide released into the atmosphere. Here in the United States, transportation releases about 33 percent of all the carbon dioxide that we produce. Perhaps even more alarming, car use is skyrocketing in the world's two most populous countries, China and India. It's easy to see that reducing or, better yet, eliminating those gas-burning engines will play a huge role in our planet's future.

But how?

Every day, traffic jams throughout the world waste millions of gallons of gasoline while pumping millions of tons of carbon dioxide into the atmosphere.

John Kohler

DATA FILE 1:
Cleaner Cars

In 1975, Congress passed CAFE, the Corporate Average Fuel Economy standards. Its intent was to reduce our nation's oil consumption by making cars and trucks more fuel-efficient. From 1975 to 1985, CAFE raised average car mileage from about 13.5 miles per gallon (mpg) to more than 27 mpg. Light truck fuel efficiency improved from 11.6 mpg to 19.5 mpg. Beginning in the 1980s, however, the automobile industry lobbied the Reagan administration to actually reduce fuel efficiency. Improvements in mileage stagnated. Congress and recent presidents added to the problem by providing huge tax breaks for small businesses to buy large, gas-guzzling trucks and sport utility vehicles, or SUVs. This encouraged millions of Americans to buy larger, more wasteful vehicles.

As concerns over global warming have skyrocketed, however, so have demands to once again increase car fuel efficiency. Increasing fuel efficiency, in fact, may be one of our fastest ways to stop the rise in CO_2 emissions. According to the Union of Concerned Scientists, CAFE standards established in 1975 saved Americans $92 billion in fuel costs in the year 2000, while keeping 720 million tons of greenhouse gases out of our atmosphere. New cars clearly show that huge, additional improvements can be made.

In 1997, Toyota announced a new car called the Prius, powered from a combination of batteries and gasoline. By 2009, dozens of models of hybrid vehicles were hitting the roads. Hybrids utilize sophisticated technology to conserve power. At slow speeds, for instance, where gas engines aren't efficient, many hybrids run only on battery-fed electric engines. At higher speeds, gas engines kick in to power the cars and recharge the batteries. By combining technologies, hybrids can achieve impressive fuel efficiencies. The best-rated hybrids, the Toyota Prius and the Honda Civic Hybrid, get about 45 miles per gallon, far higher than CAFE standards.

John Kohler

Smaller hybrid vehicles offer outstanding emissions reduction over traditional gasoline-powered vehicles. Unfortunately, many manufacturers are using hybrid technology to continue selling over-powered gas-guzzling SUVs and trucks.

Are All Hybrids the Same?

Just because a car is a hybrid doesn't mean it's necessarily good for the environment. While the Toyota Prius and Honda Civic Hybrid get great gas mileage, car manufacturers are also using hybrid technology to boost power instead of improving fuel efficiency. A 2005 Honda Accord Hybrid, for instance, got only 2 miles per gallon better mileage than a similar Accord with a V-6 all-gasoline engine. Other car manufacturers, from Ford and General Motors to Toyota and Nissan, are also trying to keep selling big—and wasteful—cars by fitting them with hybrid technology. A 2008 Toyota Tacoma Hybrid SUV, for instance, gets only 19 mpg in the city and 25 on the highway. A Chevy Tahoe Hybrid gets only 21 and 22 mpg respectively. So why are the car companies sacrificing good gas mileage for power? Easy. Car companies earn much bigger profits from large cars than from smaller ones. And they figure that to an SUV driver who is used to getting 12 or 14 miles per gallon, 20 or 21 mpg might sound pretty good. Unfortunately, hybrid SUVs and trucks help blind consumers to the real, deeper changes we need to make to protect our planet—driving smaller cars or, where possible, getting rid of cars altogether.

Toyota Prius Powertrain Schematic

Smaller electric motor used for starting and controlling engine speed

Generator

Battery

Gas Engine

Electric Motor

Wheels

Planetary gear set (allows for individual speed control)

Final Drive

The Toyota Prius has a complex hybrid system known as a dual-mode hybrid. In the Prius, a planetary gear set allows for the engine and electric motor to synergistically (in parallel) drive the wheels, or the electric motor to individually drive the wheels (with either the engine on or off). Power to the wheels can flow directly from the electric motor (for low speed/load all-electric operation), can come directly from the gasoline engine, or can be a combination of gasoline and electric motor contributions. During decelerations the electric motor is used to charge the battery pack. Energy from the gasoline engine can also be used to power the generator and charge the battery.

But even with improved gas mileage, hybrid vehicles still burn gasoline and contribute to global warming. All-electric "zero emissions" vehicles would solve that problem. Batteries in an all-electric vehicle could be charged up nightly using electricity generated by clean energy sources, such as wind power. No all-electric vehicles are currently in full-scale production in the United States, but more than a dozen small companies are building electric cars worldwide. Chevrolet has promised to release an all-electric, long-range vehicle by 2011. This car may help bring all-electric vehicles into mainstream American life.

In May of 2009, President Obama gave a boost to building better vehicles by announcing that car makers must raise U.S. fuel efficiency standards by 30 percent overall by the year 2016. To truly combat global warming, however, the world needs to begin moving beyond cars of all kinds.

DATA FILE 2:
Fewer Cars

As good as zero-emissions vehicles sound, they still add a huge burden to our environment, both directly and indirectly. To evaluate the impact of personal vehicles on global warming, we also have to consider other activities besides actually driving the car or truck around. These include:

- The amount of energy—and emissions produced—during a car's manufacture.
- The emissions produced to build roads, bridges, tunnels, parking lots, and other infrastructure necessary to support the vehicles.
- The productive land that is lost in designing cities around cars.

Building a new car requires an enormous amount of energy. Mining or acquiring the steel, aluminum, and other raw materials for the new car; refining those materials; assembling the materials; and transporting them all generate greenhouse gases. According to different estimates, the energy needed to make a car is about 10-20 percent of the energy burned up in fuel during the car's lifetime.

Perhaps an even bigger impact of cars comes from how we've designed our world around them. In cities designed for automobile transport, as much as 50 percent of the land is devoted to roads, parking lots, and other car-related spaces. The concrete needed for this infrastructure is by itself a major source of greenhouse gases. Estimates are that the concrete industry produces 7-8 percent of the world's greenhouse gas emissions.

Roads, parking lots, and other automobile infrastructure devour vast areas of land needed for agriculture, homes, watershed, and wildlife.

Land that goes toward supporting cars is also lost for agriculture, recreation, more efficient transport, and living plants that reduce carbon dioxide in the atmosphere. Unlike automobiles, plants actually consume carbon dioxide, so every acre devoted to automobile transport inflicts a double-whammy on our environment: more greenhouse gases are added to the atmosphere and fewer greenhouse gases are taken away.

Cars have also contributed to the problem of urban sprawl. Communities that are dependent on cars are more spread out. Houses, stores, and other businesses are all built farther apart and spread in many directions. This creates a vicious cycle where people have to drive more just to accomplish basic tasks. It also makes it impossible to provide people with light rail and other more efficient transportation, because there are simply too many places that have to be reached.

CONCLUSION:

When you add up all of the car's impacts on global warming and the environment, it quickly becomes clear that the automobile is the least efficient, most harmful form of transportation for our modern world. Fortunately, there are better alternatives than cars to get us where we are going. A train car carrying 22 people, for instance, uses only about half of the energy per person as an automobile with a single occupant. A bus produces similar energy savings. Furthermore, a typical light rail line can carry 15 times more people than a lane of highway, greatly reducing the land area needed for transportation corridors. Walking and bicycling can also become cornerstones to cities that are more efficient, safer, and a lot more enjoyable to live in.

The question is—How do we redesign our car-dominated cities to become more efficient?

Lynn Havens

Even sprawling cities, such as Minneapolis, Minnesota, have built light rail systems to stem automobile growth in recent years.

NEW YORK CITY HAS THE HIGHEST

rate of public transportation use in the United States. According to a 2004 report by the U.S. Census Bureau, "Of the approximately 6.4 million people nationwide who usually travel to work using public transportation, nearly one-third lives in New York City." In fact, New York is the only U.S. city where more than half of the workforce—55 percent—commutes to work by public transportation. In most of the rest of the country, only about 5 percent of commuters use public transportation while three-quarters of workers drive to work in their own cars—alone.

THE ISSUE:

According to the United Nations Population Fund, by 2008, more than half of the world's population lived in cities. This was the first time in human history that more people lived in cities than lived in rural areas. Unfortunately, most of the world's cities are hugely inefficient. Poor transportation, cheap construction, and other wasteful energy practices contribute to enormous, unnecessary production of greenhouse gases. Better planning and construction techniques hold the promise of reducing this wastefulness—improving life for urban dwellers and making a real dent in global warming.

Although not as extensive as New York's rail system, the Washington D.C. area Metro system allows easy access to many areas of our capital and surrounding communities.

BACKGROUND FILE:

With more than eight million people, New York City is the largest city in the United States. In terms of transportation, it is also the most efficient.

Why?

Looking at New York City, two obvious features lend themselves to public transportation:

1) New York City was built before the age of the automobile. In other words, the city had to build other, efficient ways for people to get around.

2) The city core is restricted by its geography—in this case, the island of Manhattan—and is forced to make good use of all available space.

Most major U.S. cities that have significant public transportation today share one or both of these features. In most newer cities, with room to spread out in all directions, fewer than 10 percent of all workers use public transportation. These sprawling cities remain car-dependent—and contribute hugely to the planet's burden of greenhouse gases.

It doesn't have to be that way.

DATA FILE 1:
Portland, Oregon—Reinventing Itself

By the 1970s, Portland, Oregon, had experienced the same sad fate as many other large American cities. Manufacturing had dried up. Many middle-class families had abandoned downtown for quieter suburbs on the city's outskirts. Malls replaced downtown stores and theaters as places for shopping and recreation. Streetcar, bus, and trolley lines had been torn up to make room for more freeways and roads. In short, the city was dying.

Over the next decades, however, the "City of Roses" began reinventing itself. It undertook a variety of projects to breathe life back into the city. These included everything from fixing up old houses and repaving streets to adding new parks and mixed-use areas where businesses and housing stood next to each other.

A key to Portland's rebirth was its commitment to rail-based mass transit. In 1978, the city took the bold step of canceling a proposed freeway that would have gutted many of the city's neighborhoods. Instead, the city decided to invest in a new light rail system. Today, the MAX light rail system consists of three lines, covering 44 miles. An 8-mile streetcar loop also connects to the MAX system, and extensive rail additions are planned. Along with mass transit, Portland has invested heavily in parks and in bike and walking trails.

Portland has also encouraged growth around its new light rail systems. All over the city, one sees new condos,

Portland's MAX light rail system is part of an integrated public transportation system that also includes buses and trolleys.

apartments, and shopping areas near MAX stations. Meanwhile, Portland has tried to limit the number of sprawling subdivisions growing up at the city's edges. This strategy is redistributing the city's population so it can be better served by mass transit instead of automobiles.

Portland still has significant traffic problems. Instead of a decaying urban core, however, most areas of the city have once again become vibrant neighborhoods. Unlike most American cities, Portland's inner-city population grew rapidly between the years 1980 and 2000, expanding about as fast as its suburbs. One quarter of its workers commute by mass transit, bicycle, or walking. More important, the city is not just a place to work, but a desirable place to live. Shopping and entertainment are within walking distance of most Portlanders, and the city is regularly ranked as the "greenest" city in the United States. The city has proven that we Americans do not have to abandon our cities—and our futures—to the automobile.

DATA FILE 2:
Getting on Board

As logical as mass transit and urban redevelopment are, they face serious hurdles. One is that federal funding for mass transit is still dwarfed by spending on roads and highways. Another is that as much as Americans complain about traffic, most of us still do not recognize the need to restructure our cities so that mass transit becomes easy and convenient to use. Traffic-clogged Vancouver, Washington, is a good example. It sits right across the river from Portland, Oregon, but has for years refused to link up with Portland's MAX system to create a convenient region-wide transit system.

Despite these obstacles, mass transit is beginning to blossom even in cities without good comprehensive growth strategies. In city after city and state after state, voters have chosen to spend money on light rail and other mass transit. According to the group Light Rail Now, by 2010,

light rail systems were operating in 30 American cities—including many of the most sprawling cities in the nation. Twenty-two cities had "heavy" or commuter rail systems operating. In the third quarter of 2008, Americans took more than 2.8 billion trips on public transportation—an increase of 6.5 percent over 2007. At the same time, vehicle miles on the nation's highways decreased by 4.6 percent.

Rail-based mass transit has been so successful that cities are scrambling to build more rail lines. These lines have an additional benefit of stimulating development and boosting regional economies. Dallas, one of the nation's most sprawling cities, plans to more than double its DART light rail network by 2013, to 91 miles of service along five different routes. As in Portland, rail commuter lines everywhere are attracting new high-density housing and business development, helping to make cities more efficient. This cuts down on wasteful urban sprawl and greenhouse gas emissions.

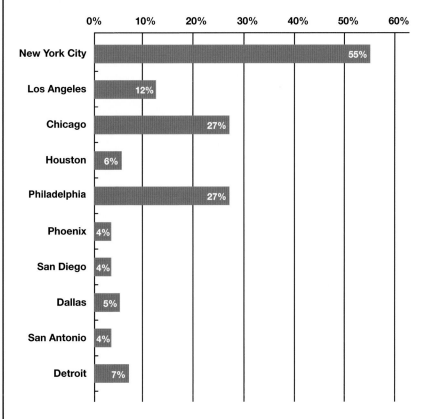

America's Top 10 Largest Cities: Percent of Workers Who Use Public Transportation

- New York City 55%
- Los Angeles 12%
- Chicago 27%
- Houston 6%
- Philadelphia 27%
- Phoenix 4%
- San Diego 4%
- Dallas 5%
- San Antonio 4%
- Detroit 7%

DATA FILE 3:
Buildings That Work

Improving transportation is just one step to "greening up" our cities. Another key is improving the energy efficiency of buildings. Buildings devour 40 percent of the energy used around the world each year. Lighting alone consumes 20 percent of the world's electricity. In the face of such numbers, people everywhere are working to create "greener" buildings that use less energy.

Many "green" changes are extremely easy and inexpensive. In South Africa, simply adding a ceiling beneath a roof has reduced 50 percent of the heat loss from some buildings. In the United States, using outside air to cool office buildings at night has reduced air-conditioning costs by 55 percent or more. Planting trees around buildings and even painting rooftops and walls so they reflect heat better can also lead to substantial energy savings. Some experts believe that improved lighting, insulation, and other features can reduce energy use in modern buildings by as much as 80 percent.

Buildings, though, can do more than save energy. They can produce energy and become a significant resource for our planet. In the past several years, large retailers have begun installing solar panels on top of their "big box" stores. Kohl's, Safeway, Whole Foods Market, and even Wal-Mart have turned hundreds of acres of wasted rooftops into power plants. The solar panels are expensive, but companies believe they can bring down costs as they install more of them. Meanwhile, an array of panels can produce 10-40 percent of the electricity used by a store. If the trend continues, these rooftop panels will help curtail the need for new coal-fired power plants and keep millions of tons of carbon dioxide out of the air.

Companies, non-profit groups, urban planners, and individuals are also experimenting with "green roofs." Green roofs are those that have a layer of living plants growing on top of them. In their simplest forms, such as the roof on one new Apple Corporation store in Chicago, a layer of one kind of green plant covers the roof like a lawn. In more advanced green roofs, people have planted prairies, trees, and even forage for farm animals.

Every green roof has a waterproof layer to keep moisture out of the building, along with a soil layer of varying depths. Even a simple green roof reaps large environmental harvests.

These include:
- reducing heating and cooling costs.
- preventing surges of storm-water runoff into sewage and river systems.
- gobbling up CO_2 from the atmosphere—and producing oxygen.
- keeping harmful sunlight from breaking down roofing materials as quickly.
- keeping rooftop air—and surrounding areas—cooler.

Some green roofs even provide habitat for wildlife and native plants. The nonprofit group Ducks Unlimited "planted" a green roof on their national headquarters building that consisted of native prairie plants. Each year, they do a controlled burn on the roof to simulate fire conditions on a natural prairie!

Linda Ferreira

Workers install a green roof on top of the new Education Building at the University of Central Michigan.

"Building a 'Green' Home"
By Rick Stern

A few years ago, with dreams of eco-building in my head, I began my "E-home" project by purchasing ten acres of forested hillside in northwestern Montana just 2½ miles from the residential school at which I teach. In 2006, while I was living on campus, a crew started building my barn (the "nouveau Redneck" name for a two-car garage). We used wood that was cut from my land and milled on site. Each day I'd ride my bike over to lend a hand after work. The next year, the same outfit that built the barn began construction of our "straw bale" house.

The "straw bale" house was built from natural materials, such as straw, wood, clay, natural dyes, and even milk!

Although it sits high on a specially designed concrete and cinder block slab, the house seems to emerge naturally from the surrounding land, because its materials are mostly of the land. The walls are filled with straw purchased from a friend about 50 miles away. The wood for the framing was cut and milled within 100 miles. The walls, inside and out, are stuccoed with clay, mined on a neighbor's place a mile away. There's even organic milk in the final layer of interior plaster, as well as a natural green dye in the finish of the upstairs exterior walls. In many ways, it feels like an adobe house. The wave and warp contours on the exterior walls make it look and feel somewhat alive.

Other e-conscious features include six photovoltaic (PV), electricity-generating solar panels on the roof which feed into the local power grid. That means, when I'm generating more power than I'm using, I'm pumping it back into the grid to be used by other people instead of just losing it. In addition, I have a few batteries I keep charged by the PV panels. They provide me with outlets and lights on those rare occasions when we lose power.

The other main "green" feature of my house is its passive solar design. A large concrete slab in the interior of the house absorbs heat from the sun (when it shines), stores it, and then radiates it back into the air when the house cools off. In addition, because I live on ten wooded acres of land in a part of Montana without access to natural gas, I use a wood stove with a fan to provide additional heat. The "Ecofan"

The concrete floor absorbs heat from the sun (on sunny days), stores it, and then radiates it back into the air when the sun goes down.

← Ecofan

The woodstove is fueled by wood from trees on my own land, and then the heat from the stove is circulated by an ecofan which uses no electricity.

creates its own electricity from the heat of the stove. By using a fan to disperse the heat, there is a large increase in the wood stove's efficiency. Once thermal curtains cover the double-paned windows, the heat loss should be minimal. Finally, I have a big propane tank that fuels my energy-efficient, "on-demand" water heater and the back-up heaters I use during the coldest part of the winter.

Even though I already conserve water, I'm working on a system to collect and store rainwater. In the future when I have planted my gardens, this will ease the pressure on my well to provide for irrigation.

Will these eco-friendly, green projects ever end? I doubt it, and I hope not. As spring arrives in my corner of Montana, I am looking forward to moving from my inside tasks to creating gardens and other outdoor projects to enhance and improve this patch of land I call home.

DATA FILE 4:
Other City Success Stories

Forty years ago, the island city-nation of Singapore was a ramshackle maze of densely populated villages and shantytowns. Auto and truck traffic had almost paralyzed the city's roads, creating horrible pollution and curtailing economic growth. Beginning in the 1970s, the nation launched a massive makeover. It began building housing and businesses along corridors that could easily be served by subway and light rail lines. It also reconfigured the city into distinct "satellite" districts. As much as possible, each district was to be self-sufficient, meaning that people would not have to travel to the city's central core for their needs. Today, more than half of Singapore residents can walk to a mass transit station, while others can easily reach one by bus. This new efficiency has helped Singapore become one of the world's most modern and wealthiest nations.

Other cities around the world offer examples of extraordinary progress. In 2000, Barcelona, Spain, set out to reduce its production of greenhouse gases by 20 percent through a program of improving energy efficiency and taking advantage of Spain's abundant sunshine. The cornerstone to the plan was to require that all new or retrofitted buildings use solar energy to heat their hot water. Solar hot water systems are some of the cheapest solar systems to install, but highly effective. This ordinance alone has reduced the amount of CO_2 the city produces by 2,170 tons (1973 metric tons) per year. By 2006, more than seventy other Spanish cities had followed Barcelona's lead.

Jeremy Carver

Locals purchase tickets for Singapore's Mass Rapid Transit system (MRT). The system currently has 70 stations with nearly 74 miles (119 Kilometers) of line.

CONCLUSION:

The potential for cities to help solve global warming and other environmental problems is virtually unlimited. Programs to transform cities, however, depend on government support and the insistence of their citizens to make changes. One reason American corporations have started adding solar panels to their roofs is that they receive government tax breaks to do this. Without these tax incentives, progress will be made much more slowly—or not at all. Communicating our concerns to all levels of government (local, state, and federal) is important, but there's much more that we can do.

Sneed B. Collard III

Photovoltaic panels help provide the energy needs of this new baseball park in Midland, Michigan.

CHAPTER SIX
MAKING PERSONAL DECISIONS

Shopping at farmers' markets offers a great—and fun—way to save energy and support local economies.

IT'S PROBABLY CLEAR TO YOU BY NOW

that global warming is big and complex. It may be the most complex thing you've ever had to think about. A strong temptation when faced with such an issue might be to just throw up your hands and say, "Forget it. I can't deal with this." I feel that way sometimes.

But remember this. While it's true that you may not be able to build a solar power plant by yourself or shift the federal budget away from building new freeways, your ability to improve the situation is greater than almost anyone else's on the planet. Why? Because we Americans have more resources at our disposal than almost anyone else. We also consume more, drive more, and waste more than almost anyone else. Each American is responsible for producing four times the CO_2 as each person in China; ten times the CO_2 as each person in India; and twice as much as each person in Japan, Iran, Italy, and the United Kingdom. This means that our potential to make a difference is that much greater.

But how does a person even begin? My own approach is to build my efforts like a sustainable "green" building—from the foundation to the rooftop.

THE FOUNDATION:
Smarter Habits

Personal habits can greatly reduce your "carbon footprint" on the planet. These require no more effort than learning any other good habit, such as opening a door for someone or remembering to take the trash out every Thursday morning. Things I focus on especially include:

1) Remembering to turn out the lights when I leave the room.
2) In winter, wearing warmer clothes indoors instead of cranking up the heat to tropical levels.
3) Remembering to bring my own bags when shopping.
4) Combining my errands, so that I'm making fewer trips in my car.
5) Turning off my computer and appliances at night.
6) Turning down the heat at night and whenever I leave the house.
7) Opening and closing window shades to light, heat, or cool the house instead of using artificial lights, heaters, and air conditioners.
8) Reusing drinking glasses several times before putting them in the dishwasher.

These simple habits alone can cut a person's energy use—and "personal" greenhouse gas emissions—by half. They also save you and/or your parents a great deal of money in utility and gasoline costs.

FIRST FLOOR:
Cheap Changes

Beyond my personal habits, I've also invested in simple changes around the house that have dramatically reduced my own carbon footprint.

Fluorescent light bulbs

Incandescent light bulbs account for 20 percent of the electricity used in homes in the United States. According

Taking your own bags when you shop saves energy by eliminating the manufacture of new plastic and paper bags.

Sneed B. Collard III

Lynn Havens

Replacing incandescent bulbs with efficient fluourescents is one of the easiest and most dramatic ways to decrease our individual "carbon footprints."

electric pencil sharpeners—continue to consume power even when they are turned off. By plugging these devices into a high-quality power strip, however, you can just turn the power strip off at the end of the day, and completely shut down the power-gobbling devices.

Weatherstripping

Houses and other buildings lose huge amounts of heat—or cold—because they are not well insulated or tightly sealed. Insulation in walls and ceilings can cost a lot—as can better windows and doors. Drafty doors and windows, however, can be greatly improved with cheap weatherstripping from any hardware store. This could save your family money on heating and cooling costs—and, again, cut down on your CO_2 share. Many local energy companies even provide these materials for free.

to the Environmental Protection Agency (EPA), if every household replaced one incandescent bulb with a fluorescent light bulb, it would save enough energy to power three million homes. Fluorescent bulbs do cost more than regular bulbs at first, but they are getting cheaper all the time, and they last much longer. In the four years I've lived in my current house, only two fluorescent bulbs have blown out. Some people have a concern that

fluorescent bulbs contain poisonous mercury, and it's true that they do. However, according to the EPA and U.S. Department of Energy, that mercury is far less than the mercury released by a coal-fired power plant producing the extra energy needed for regular old-style bulbs.

Power strips

Many of today's appliances—televisions, DVD players, phones, printers, even

SECOND FLOOR:
Developing A "Green" Mindset

The more I understand global warming's causes, the more deeply I look at every aspect of my life to see how I can reduce my impact on the planet.

Beyond recycling

We've all been bombarded with messages to recycle and, no doubt, recycling is important. However, to drink a bottle of water and then put the bottle in the recycling bin does not prevent that bottle from having to be made in the first place. That takes energy. It's far more important to reduce the amount of stuff we consume from the beginning. Instead of buying bottles of drinking water, for instance, I carry my own water bottle with me and fill it up from taps and drinking fountains. When I get a take-out meal, instead of taking plastic spoons and forks from the restaurant, I take my food home and use my own utensils. There are many other ways to reduce our consumption. These can be as simple as wearing a pair of shoes for an extra month or two before buying a new pair or keeping your sunglasses in a case, so you don't damage them and have to buy new ones. Over time, these little changes add up in a big way.

Buying things that last

Part of reducing consumption is to buy quality goods and use them for a long time. I drive a car that is twenty-four years old. It still gets reasonable gas mileage and drives well. Before I bought it, I researched the reliability of different car makes in the magazine *Consumer Reports*. My research paid off. During the time I've driven my one car, most people I know have bought and sold four or five different cars. My choice has not only saved me tens of thousands of dollars, it has prevented thousands of tons of extra carbon dioxide from going into the atmosphere during the construction of those three or four "extra" cars. Higher quality items often cost a bit more, but they prove cheaper in the long run, because they last much longer.

The author standing next to his 24 year-old car.

Buying local

We are used to buying things as cheap as possible, but cheap things often carry a hidden energy cost. Items made in China, for instance, have to be shipped all the way across the Pacific Ocean to reach our shores, and then by rail or truck to the store near us. Each step consumes energy and produces greenhouse gases. Whenever I can, I try to choose something that's made closer to me, especially clothing, paper goods, furniture, and food. Again, these items may be a bit more expensive, but they consume far less energy in their "production life-cycles." Buying locally has the added benefit of employing American workers.

Walk, bike, or ride

Walking and biking are without a doubt the best ways to get around. They produce no greenhouse gases and give us exercise. I've also started looking for other ways to avoid using my car. This is easier in some places than others, but I do have a bus route near my house that goes downtown, and I try to use it once a week. Each car trip we can prevent—no matter who is driving—makes an important contribution to slowing global warming.

Eat less meat

Meat production consumes more water, energy, and other resources than any other form of food production on the planet. Think about eating less meat—and read "A Vegetarian Speaks."

"A Vegetarian Speaks"
by Richard Moser

I stopped eating meat a month or so after finishing high school. There's been an accidental bite here and there, but otherwise I have been a vegetarian for 30 years. That means no red or white meat, no seafood, no fish, no Thanksgiving turkey, no shrimp paste or chicken stock, nothing "with a face." I also don't eat egg-oriented foods, although if they're a minor ingredient, that's okay, plus I eat a little dairy. Everyone has to draw their lines somewhere.

I had started cleaning up my diet three years before, by eating much less junk food. Looking back on my childhood, I was frequently sick. I think it was because I ate mostly unhealthy stuff: lots of meat, dairy, and sugar. In addition, I had many cavities. It started to sink into my consciousness that eating sugar wasn't good for me, so I cut it back a lot. During my sophomore year in high school, I missed two weeks with a particularly nasty stomach flu, during which I lost ten pounds that, as a skinny kid, I wouldn't have thought possible. Improving my diet, though, brought immediate benefits. I became a better student because I wasn't sick as much. My acne started to clear up. I felt better about myself and slowly began to expand my social contacts. I wasn't so shy anymore.

For me, learning to eat differently came easily. I had a natural interest in food, and I just started to explore. When you don't have meat and milk to fall back on, you start to notice how many different foods there are out there. It's really very easy to find something other than a hamburger, you just have to first rule it out as an option. Over time, I learned how to make healthy food items, and

Tim Braun

I discovered lots of places where I could buy them.

The more I learned about health—and the more I learned about the environment, since ecology was my favorite subject in school—the more committed a vegetarian I became. A meat-based diet is not required for humans to live healthy lives. In fact, it causes many degenerative diseases, such as plaque in the arteries, which leads to heart disease and increased stroke risk. A meat-based diet also is ecologically destructive, compared to a plant-based one. It takes several pounds of grain and thousands of gallons of water to create a pound of beef, so it's not a terribly efficient use of resources to eat meat.

A meat-based diet makes no sense whatsoever in terms of social justice, as only the wealthier people of the world can afford to eat it. Animals and the processing necessary to support meat-based diets create a great deal of pollution, especially the global-warming gas methane. A meat-based diet is incompatible with having compassion for the suffering of animals, most of which are raised in stunningly inhumane and unhealthy conditions. I have spiritual reasons as well for being a vegetarian. The biggest is that in order for me to live, it is not necessary that another animate being die.

So if you want to eat meat, I say do it consciously and be honest about why you do it, because the more you learn, the less you'll want to do it.

For more information on being a vegetarian:

FDA's Food and Nutrition Information Center
    <http://fnic.nal.usda.gov (search for "vegetarian")>

Medline Plus
    <http://www.nlm.nih.gov/medlineplus/vegetariandiet.html>

Pew Commission on Industrial Farm Animal Production
    <http://www.ncifap.org/>

EarthSave
    <http://www.earthsave.org/>

THE "GREEN" ROOF:
Getting Involved

All of the previous steps greatly reduce our impacts on global warming. But there's more—much more—we can do. Many of the large decisions about energy, transportation, and conservation are made by corporations and governments. As you read earlier, the United States dragged its feet on preventing global warming for more than a decade. Pressure by ordinary citizens like you and me can make the difference in whether or not we as a society begin to phase out fossil fuels, provide more money for rail transportation, or provide tax incentives for people to insulate their houses. Here are things I do to influence these decisions:

• Write letters to the editors of my local newspapers.

• Write or phone my U.S. Congressman and U.S. Senators asking that they commit the United States to reducing greenhouse gas emissions to specific, lower levels.

• Write or phone my state representatives asking that they also take action on global warming.

• Join groups mobilized to improve United States policies and action on the issue.

A number of national and international organizations are dedicated to solving the global warming crisis. They are always looking for help. The Alliance for Climate Protection was started by former Vice President Al Gore. The alliance is a coalition of leaders from all sectors of society. One project they have started is called the We Campaign, aimed specifically at spurring government action on global warming. To get involved, look up <http://www.wecansolveit.org/>.

Another group active in the climate crisis is the Union of Concerned Scientists. It is involved in several campaigns to publicize good scientific knowledge and improve public policy. Anyone can join and get involved. The group's Web site has useful information for how to write letters to the editor, call your representatives, and other activities. Web link: <http://www.ucsusa.org>.

The Project Vote Smart site can tell you how your representatives vote on different issues. It also provides links for how to contact them. Web link: <http://www.votesmart.org/index.htm>.

Contact information for federal representatives can also be found at: <http://www.congress.org/congressorg/home/>.

You also might think about starting or joining a group at your school dedicated to taking action on global warming and sharing information with your fellow students. The more you get involved, the more difference you can make. Instead of feeling like a victim of global warming, you can rightly feel good that you are part of the solution.

Students Taking Action

Thousands of students around the country and around the world have begun projects to fight global warming and educate their schools and communities about this crisis. These projects range from making videos to starting energy consulting firms to writing plays and doing fundraising to install solar panels at schools. Here are just a few links to get some ideas:

California Climate Champions
<http://climatechamps.org/champs/>

Action For Nature "Eco-Hero"
<http://www.actionfornature.org/?tag=2010-eco-heroes>

Create Your Future
<http://www.kidsforfuture.net/index.php>

World Changing
<http://www.worldchanging.com/>

Cool Schools project
<http://www.climateprotectioncampaign.org/sonomaccp/coolschools.php>

Glossary

Alternative energy—an energy source that produces little or no pollution and does not have the environmental problems of fossil fuels or nuclear power. Most often refers to solar, wind, wave, and hydroelectric power.

Biofuels—fuels, including ethanol, that are manufactured from agricultural products, such as corn, sugar cane, and wood pulp.

Carbon dioxide (CO_2)—the major product released from burning fossil fuels or biofuels. It is the most abundant man-made greenhouse gas and a major cause of global warming.

Carbon footprint—the amount of carbon dioxide a person is responsible for creating. This results from the person's driving habits, food and clothing purchases, and use of electricity, among other things.

Climate change—long-term changes in temperature, precipitation, and weather patterns resulting from global warming.

Coral bleaching—the loss of a coral's zooxanthellae resulting from a severe stress, usually warmer ocean temperatures.

Fossil fuels—fuels consisting mostly of carbon that have been created by ancient geological processes. The main fossil fuels are oil (petroleum), coal, and natural gas.

Global warming—the rise in earth's average surface temperature.

"Green"—any quality or habit that leads to a lower impact on the earth's environment.

Greenhouse gases—gases in the atmosphere that trap heat, leading to a rise in earth's overall temperatures. The most important greenhouse gases, in order of abundance, are water vapor, carbon dioxide (CO_2), methane (CH_4), nitrous oxide, ozone, and chlorofluorocarbons, or CFCs.

Hybrid vehicles (hybrids)—cars and other vehicles that rely on more than one source of power, usually a gasoline engine and an electric motor.

Hydroelectric—electricity produced by the flow of water through electric turbines.

Ice cores—long tubes of ice drilled from glaciers that reveal changes in precipitation and the environment over long periods of time.

Industrial revolution—a period between the years 1760 and 1850 that witnessed rapid advances in farming, manufacturing, and transportation. Many of these advances were the result of new technologies, such as the steam engine and new machines to spin and weave cloth.

Light rail—electric trains that are used to transport people rapidly around urban areas. They are not as big or fast as heavier, longer-distance trains for people and freight.

Mass transit—transportation, such as light rail, commuter rail, and buses, that can be used by anyone. (See "public transportation.")

Megawatt—a unit of power equaling one million watts. A rough guide is that a megawatt can power between 500 and 1000 homes.

Methane (CH_4)—a common greenhouse gas. It is a component of natural gas, but is also produced by cows, rice cultivation, marshlands, termites, and many other human and natural processes.

Nuclear power—electricity generated from the radioactive decay of uranium or plutonium. As these fuels decay, they release huge amounts of energy. This energy is used to heat water that produces steam. The steam drives turbines that generate electricity.

Petroleum—a fossil fuel created by the breakdown of plant matter over geologic time periods. The major components of petroleum are carbon and hydrogen. Also known as "oil" or "crude oil."

Photovoltaic (PV)—referring to silicon devices that can turn sunlight directly into electric current.

Public transportation—any form of transportation available to the general public. Most often refers to buses and passenger trains.

Renewables—sources of energy that do not deplete earth's resources or significantly damage the environment. Most often refers to hydroelectric, wind, and solar power, but can also include wave power, geothermal, and other power sources. See "alternative energy."

Solar Power—any form of energy originating from the sun. See "photovoltaic" and "solar thermal."

Solar thermal—refers to using the heat from the sun as an energy source, usually to heat up water for domestic use or to drive steam-powered turbines.

Sustainable energy—see "renewables."

Wave farm—an array of machines that generate electricity from the up and down movement of ocean waves.

Wind farm—an array of windmills that generate electricity.

Zero emissions vehicle—a vehicle that does not release any pollutants, including greenhouse gases. Most often refers to all-electric vehicles.

Zooxanthellae—microscopic organisms that live inside corals, giant clams, and other marine animals. They use sunlight to make food that they share with their hosts, and they are a major source of nutrition for their hosts.

A Note on Research

The Internet, or the "Web," has revolutionized the research process in both bad and good ways. Never before has so much information been available to us so quickly. Unfortunately, a great deal of that information has never been checked for accuracy and suffers from extremely biased viewpoints.

Usually, when I research a new book, I rely on, in order: other books by experts, scientific papers published in journals, personal interviews, newspaper and magazine articles, and finally, information from the Web. The topics of energy and global warming, however, are extremely fast-moving. Almost before information is published, it is out of date. Because of this, I used the Web much more heavily than I normally would. To make sure that I used only reliable information, I focused on the following sites:

• Official government sites with official statistics. Example: U.S. Department of Transportation.

• Sites from scientific and other academic organizations with high ethical standards and a commitment to truth. Example: The Union of Concerned Scientists

• Sites from trade or advocacy organizations that keep noncontroversial statistics about energy and similar topics. Example: American Public Transportation Association

Every once in a while—if, for instance, I just needed to find out the populations of the world's largest cities—I would look it up on a more commercial site such as Wikipedia or InfoPlease, but I never used these sites for any critical information. One place the Web has really become valuable, however, is to locate news and scientific articles that have been published elsewhere in print form. I was able to find many useful articles from The *New York Times*, the Associated Press, and other reputable sources that have been posted online.

REFERENCES AND RESOURCES

BOOKS (PRINT AND ONLINE)

Allsopp, M., Page, R., Johnston, P., and Santillom D. (2007). *Oceans in Peril: Protecting Marine Biodiversity*. Worldwatch Report 174, Worldwatch Institute, Washington, D.C.

Gore, A. (2006). *An Inconvenient Truth*. Rodale, Emmaus, PA.

Mastny, L. (editor). (2010, October). Renewable Energy and Energy Efficiency in China: Current Status and Prospects for 2020. Worldwatch Report 182, Worldwatch Institute, Washington, D.C.

Nadis, S. and MacKenzie, J. J. (1993). *Car Trouble*. Beacon Press, Boston.

Newman, P. and Kenworthy, J. (2007). "Greening Urban Transportation." In State of the World 2007: *Our Urban Future*. W.W. Norton & Co., New York.

Pidwirny, M. (2006). "Causes of Climate Change." *Fundamentals of Physical Geography, 2nd Edition.* Retrieved October 15, 2009, from <http://www.physicalgeography.net/fundamentals/7y.html>

Sawin, J. L. and Hughes, K. (2007). "Energizing Cities." In *2007 State of the World: Our Urban Future*. W.W. Norton & Co., New York.

The Worldwatch Institute (2006). *Vital Signs 2006-2007*. The Worldwatch Institute and W.W. Norton & Co., New York.

MAGAZINE AND JOURNAL ARTICLES (PRINT AND ONLINE)

Bacher, J. (2002, Spring). "Petrotyranny." Earth Island Journal, Spring, 2002. Retrieved November 10, 2009, from <http://www.thirdworldtraveler.com/Oil_watch/Petrotyranny.html>

Bettelheim, A. (2006, September 29). "Biofuels Boom." CQ Researcher. 16:34.

Cayan, D. R., Bromirski, P. D., Hayhoe, K., Tyree, M., Dettinger, M. D., and Flick, R. E. (2008, March). "Climate change projections of sea level extremes along the California coast." (Report). Climate Change, 87:1, pp. 57-73.

Comras, V. (2005, January). "Al Queda Finances and Funding to Affiliated Groups." Strategic Insights, IV:1 Retrieved November 10, 2009, from <http://www.ccc.nps.navy.mil/si/2005/Jan/comrasJan05.asp>

The Economist editorial staff.(2008, November 13). "Don't Bail Out the Carmakers." The Economist. Retrieved November 10, 2009, from <http://www.economist.com/opinion/displaystory.cfm?story_id=12601932>

"Ethanol's Impact on Food Prices." (2007, May 22). Corn & Soybean Digest. Retrieved November 10, 2009, from <http://cornandsoybeandigest.com/biofuels/ethanol-impact-food-prices/>

Hoegh-Guldberg, O., Mumby, P. J., Hooten, A. J., Steneck, R. S., Greenfield, P., Gomez, E., et al. (2007, December 14). "Coral Reefs Under Rapid Climate Change and Ocean Acidification." Science, 318:5857, pp. 1737-1742.

Kaplan, D. E. (2003, December 7). "The Saudi Connection: how billions in oil money spawned a global terror network." U.S. News & World Report. Retrieved November 10, 2009, from <http://www.usnews.com/usnews/news/articles/031215/15terror.htm>

Lauchbaum, D. (2008, March). "David Lauchbaum: Comments from a Nuclear Engineer with the Union of Concerned Scientists." Nuclear News, pp. 33-37.

Leibtag, E. (2008, February). "Corn Prices Near Record High, But What About Food Costs?" Amber Waves. Retrieved November 10, 2009, from <http://www.ers.usda.gov/AmberWaves/February08/Features/CornPrices.htm>

Long, M. E. (2002, July). "Half Life." National Geographic Magazine Online Extra. Retrieved November 10, 2009, from <http://ngm.nationalgeographic.com/ngm/0207/feature1/fulltext.html>

Peterson, C. H., Rice, S. D., Short, J. W., Esler, D., Bodkin, J. L., Ballachey, B. E., Irons, D. B. (2003, December 19). "Long-Term Ecosystem Response to the Exxon Valdez Oil Spill." Science. 30:5653), pp. 2082-2086.

Pucher, J., Peng, Z-R, Mittal, N., Zhu, Y., Korattyswaroopam, N. (2007, July). "India: Impacts of Rapid Economic Growth." Transport Reviews. 27: 4, pp.379–410. Retrieved November 10, 2009, from <http://policy.rutgers.edu/faculty/pucher/PUCHER_China%20India_Urban%20Transport.pdf>

Roach, J. (2007, September 17). "Arctic Melt Opens Northwest Passage." National Geographic News. Retrieved November 10, 2009, from <http://news.nationalgeographic.com/news/2007/09/070917-northwest-passage.html>

Woynillowicz, D. (2007, September/October). "Tar Sands Fever!" World Watch. 20:5.

NEWSPAPERS (PRINT AND ONLINE)

Botkin, D. B. (2007, October 17). "Global Warming Delusions." The Wall Street Journal Asia.

Bradsher, K. (2008, April 17). "From Six-Year Drought in Australia, a Global Crisis Over Rice." New York Times.

De Souza, M. (2008, December 9). "Billions of litres of tainted oilsands water leaking: Report." Canwest News Service. Retrieved November 10, 2009, from <http://oilsandstruth.org/billions-litres-tainted-tar-sands-water-leaking-report>

Doyle, L. (2008, July 4). "Storm Over Cape Cod." The Independent. Retrieved November 10, 2009, from <http://www.independent.co.uk/news/world/americas/storm-over-cape-cod-860004.html>

Galbraith, K. (2009, February 17). "Obama Signs Stimulus Packed with Clean Energy Provisions." New York Times. Retrieved September 16, 2010, from <http://green.blogs.nytimes.com/2009/02/17/obama-signs-stimulus-packed-with-clean-energy-provisions/>

"Global Warming Causing Disease to Rise." (2006, November 14). Associated Press. Retrieved November 10, 2009, from <http://www.msnbc.msn.com/id/15717706/>

"History of Global Warming." Global Warming Newspaper Archive. Retrieved November 10, 2009, from <http://www.globalwarmingarchive.com/History.aspx>

Jha, A. (2008, September 25). "Making Waves: UK Firm Harnesses Power of the Sea . . . in Portugal." The Guardian. Retrieved November 10, 2009, from <http://www.guardian.co.uk/technology/2008/sep/25/greentech.alternativeenergy>

Kissinger, H. A. (2007, January 18). "The New Iraq Strategy." International Herald Tribune. Retrieved November 10, 2009, from <http://zfacts.com/p/711.html>

Landler, M. (2008, May 16). "Solar Valley Rises in an Overcast Land." The New York Times

Lean, G. (2006, December 24). "Disappearing World: Global Warming Claims Tropical Island." The Independent. Retrieved November 10, 2009, from <http://www.independent.co.uk/environment/climate-change/disappearing-world-global-warming-claims-tropical-island-429764.html>

Macartney, J. (2007, September 27). "Three Gorges Dam Is a Disaster in the Making, China Admits." The Times. Retrieved November 10, 2009, from <http://www.timesonline.co.uk/tol/news/world/article2537279.ece>

Marks, K. (2006, October 25). "Rising Tide of Global Warming Threatens Pacific Island States." The Independent. Retrieved November 10, 2009, from <http://www.independent.co.uk/environment/climate-change/rising-tide-of-global-warming-threatens-pacific-island-states-421493.html>

Murawski, J. (2006, February 21). "Putting a Price on Nuclear Power." The News and Observer. Retrieved November 10, 2009, from <http://www.redorbit.com/news/science/398986/putting_a_price_on_nuclear_power/index.html?source=r_science>

Rosenbloom, S. (2008, August 10). "Giant Retailers Look to Sun for Energy Savings." New York Times. Retrieved November 10, 2009, from <http://www.nytimes.com/2008/08/11/business/11solar.html>

Shasha, D. (Editor). (2009, April 17). "Three Gorges Reservoir Sees 166 Geo-disasters Since Last September." Xinhua News Agency. Retrieved November 10, 2009, from <http://news.xinhuanet.com/english/2009-04/17/content_11200322.htm>

"Study Confirms Greenland Ice Sheet Melt Due to Global Warming." (2008, January 16). CORDIS News. Retrieved November 10, 2009, from <http://cordis.europa.eu/fetch?CALLER=EN_NEWS&ACTION=D&SESSION=&RCN=28998>

Webb, T. (2010, September 13). "Gulf oil spill: compensation should cost less than $20bn – Bob Dudley." The Guardian. Retrieved September 15, 2010, from http://www.guardian.co.uk/business/2010/sep/13/gulf-oil-spill-bp-dudley>

Yardley, J. (2007, November 19). "Chinese Dam Projects Criticized for their Human Costs." New York Times. Retrieved November 10, 2009, from <http://www.nytimes.com/2007/11/19/world/asia/19dam.html?pagewanted=1&_r=1>

ONLINE REPORTS AND RESOURCES

"The 1960s, a Time of Turbulence and Economic Uncertainty." Economic Development Board, Singapore. Retrieved November 9, 2009, from <http://www.edb.gov.sg/edb/sg/en_uk/index/about_edb/our_history/the_1960s.html>

"2008 among 10 warmest years on record, UN reports." (2008, December 17). UN News Centre. Retrieved November 9, 2009, from <http://www.un.org/apps/news/story.asp?NewsID=29342&Cr=climate&Cr1=>

"400 Richest Americans." (2007, September 20). Forbes Retrieved November 9, 2009, from <http://www.forbes.com/lists/2007/54/richlist07_The-400-Richest-Americans_Rank_5.html>

Bellis, M. "The History of the Automobile." About.com Inventors. Retrieved November 9, 2009, from <http://inventors.about.com/library/weekly/aacarssteama.htm>

Bernstein, L., Bosch, P., Canziani, O., Chen, Z., Christ, R., Davidson, O., et al. (2007, November 17). "Climate Change 2007: Synthesis Report; Summary for Policymakers." Intergovernmental Panel on Climate Change. Retrieved November 9, 2009, from <http://www.ipcc.ch/pdf/assessment-report/ar4/syr/ar4_syr_spm.pdf>

"Blown-out BP well finally killed at bottom of Gulf." (2010, September 19). Yahoo News. Retrieved September 19, 2010, from <http://news.yahoo.com/s/ap/20100919/ap_on_bi_ge/us_gulf_oil_spill>

Burns, S. (2005, April 23). "Extra: SUV owners get free gas—courtesy of Uncle Sam." MSN Money. Retrieved November 9, 2009, from <http://moneycentral.msn.com/content/invest/extra/P115791.asp>

"Chevrolet Volt." Chevrolet. Retrieved November 9, 2009, from <http://www.chevrolet.com/electriccar/?evar10=DIVISIONAL_FUEL_EFFICIENCY>

"City of Barcelona, Spain." International Council for Local Environmental Initiatives (ICLEI)—Local Governments for Sustainability. Retrieved November 9, 2009, from <http://www.iclei.org/index.php?id=1156>

"Clean Coal Marketing Campaign." (2009, September 14). SourceWatch. Retrieved November 9, 2009, from <http://www.sourcewatch.org/index.php?title=Clean_Coal_Marketing_Campaign>

"Clean Vehicles—Cars, Trucks & Air Pollution." Union of Concerned Scientists. Retrieved November 9, 2009, from <http://www.ucsusa.org/clean_vehicles/vehicles_health/cars-trucks-air-pollution.html>

"Climate Change—Basic Information." EPA climate change. Retrieved November 9, 2009, from <http://www.epa.gov/climatechange/basicinfo.html>

"Countries with Highest GDP per Capita." aneki.com. Retrieved November 9, 2009, from <http://www.aneki.com/countries_gdp_per_capita.html>

"Deepwater Horizon/BP Oil Spill: Size and Percent Coverage of Fishing Area Closures Due to BP Oil Spill." (2010, September 13). NOAA Fisheries Service. Retrieved September 13, 2010, from <http://sero.nmfs.noaa.gov/ClosureSizeandPercentCoverage.htm>

"Deepwater Horizon oil spill." (2010, August 26). The Encyclopedia of Earth. Retrieved September 15, 2010, from <http://www.eoearth.org/article/Deepwater_Horizon_oil_spill>

EarthRights International, Racimos de Ungurahui, Amazon Watch. (2007). "A Legacy of Harm: Occidental Petroleum in Indigenous Territory in the Peruvian Amazon." Retrieved November 8, 2009, from <http://www.earthrights.org/files/Reports/A_Legacy_of_Harm.pdf>

Eckerson, C., Jr. (2006). "Defeat of the Mt. Hood Freeway." (Video). Retrieved November 9, 2009, from <http://portlandtransport.com/archives/2006/06/the_road_freewa.html>

"Emissions of Greenhouse Gases Report." (2008, December 3). Energy Information Administration. Retrieved November 9, 2009, from <http://www.eia.doe.gov/oiaf/1605/ggrpt/carbon.html>

"Energy Required to Manufacture Typical Vehicle." Google Answers. Retrieved November 9, 2009, from <http://answers.google.com/answers/threadview?id=433981>

"Environment Groups to Senate: Reject Amendments Promoting More Subsidies for Nuclear Power in Proposed Climate Bill." (2008, June 4). Union of Concerned Scientists, press release. Retrieved November 9, 2009, from <http://www.ucsusa.org/news/press_release/environment-groups-to-senate-0121.html>

"Environmental Impact: Cement Production and the CO2 Challenge." Ecosmart. Retrieved November 9, 2009, from <http://www.ecosmartconcrete.com/enviro_cement.cfm>

"Ethanol's Impact on Food Prices." (2007, May 22). Corn & Soybean Digest. Retrieved November 9, 2009, from <http://cornandsoybeandigest.com/biofuels/ethanol-impact-food-prices/>

"Findings of the IPCC Fourth Assessment Report: Climate Change Impacts." (2007). Union of Concerned Scientists . Retrieved November 9, 2009, from <http://www.ucsusa.org/global_warming/science_and_impacts/science/findings-of-the-ipcc-fourth.html>

"Fuel Economy Basics." (2007, November 26). Union of Concerned Scientists. Retrieved November 9, 2009, from <http://www.ucsusa.org/clean_vehicles/solutions/cleaner_cars_pickups_and_suvs/fuel-economy-basics.html>

"Fueling Terror." (2004). Institute for the Analysis of Global Security. Retrieved November 9, 2009, from <http://www.iags.org/fuelingterror.html>

Gable C. and Gable, S. "Hybrid Cars and Alt Fuels." About.com Hybrid Cars and Alt Fuels . Retrieved November 9, 2009, from <http://alternativefuels.about.com/od/electricvehicles/tp/2008-Electric-Vehicles.htm>

"Gasoline Explained–Where Our Gasoline Comes From." Energy Information Administration. Retrieved November 9, 2009, from <http://www.eia.doe.gov/bookshelf/brochures/gasoline/index.html>

"Global Wind Installations Boom, Up 31% in 2009." (2010, February 4). Renewable Energy World. Retrieved September 15, 2010, from <http://www.renewableenergyworld.com/rea/news/article/2010/02/global-wind-installations-boom-up-31-in-2009>

"Good Practice Case Study: Sustainable Energy Solutions in Barcelona-Spain." (2007). European Commission Directorate-General for Energy and Transport. Retrieved November 9, 2009, from <http://www.managenergy.net/products/R1574.htm>

Handwerk, B. (2005, August 8). "Hybrid Cars Losing Efficiency, Adding Oomph." National Geographic News. Retrieved November 9, 2009, from <http://news.nationalgeographic.com/news/2005/08/0808_050808_hybrid_cars.html>

Hargreaves, S. (2008, July 15). "U.S. Gas: So Cheap It Hurts." CNNMoney.com. Retrieved November 9, 2009, from <http://money.cnn.com/2008/05/01/news/international/usgas_price/?postversion=2008050109>

"History of Fuel Economy: One Decade of Innovation, Two Decades of Inaction." Pew Research Paper on the Pew Campaign for Fuel Efficiency. Retrieved November 9, 2009, from <http://www.pewfuelefficiency.org/docs/cafe_history.pdf>

"History of Kyoto Protocol." Pew Center on Global Climate Change. Retrieved November 9, 2009, from <http://www.pewclimate.org/history_of_kyoto.cfm>

"History of Rail Transport." Wikipedia. Retrieved November 9, 2009, from <http://en.wikipedia.org/wiki/History_of_rail_transport>

"How Ethanol is Made." Renewable Fuels Association. Retrieved November 9, 2009, from <http://www.ethanolrfa.org/resource/made/>

"Hybrid Vehicles—Compare Side-by-Side." U.S. Department of Energy. Retrieved November 9, 2009, from <http://www.fueleconomy.gov/feg/hybrid_sbs.shtml>

"Hydrogen Fuel Cells." (2008, November). U.S. Department of Energy Fact Sheet. Retrieved November 9, 2009, from <http://www1.eere.energy.gov/hydrogenandfuelcells/pdfs/doe_h2_fuelcell_factsheet.pdf>

"Hydrogen Production." (2008, November). U.S. Department of Energy Fact Sheet. Retrieved November 9, 2009, from <http://www1.eere.energy.gov/hydrogenandfuelcells/pdfs/doe_h2_production.pdf>

"Information on Compact Fluorescent Light Bulbs (CFLs) and Mercury." (2008, July). Environmental Protection Agency: EnergyStar. Retrieved November 9, 2009, from <http://www.energystar.gov/ia/partners/promotions/change_light/downloads/Fact_Sheet_Mercury.pdf>

Kasotia, P. (2008). "Ethanol from Corn: A Solution to Oil Dependence?" UN Chronicle. Retrieved November 9, 2009, from <http://www.un.org/Pubs/chronicle/2008/webarticles/080111_ethanol.html>

Kenworthy, J. (2000). "The Singapore/Hong Kong Success Stories and Their Implications for Developing Cities." Murdoch University Institute for Sustainability and Technology Policy. Retrieved November 9, 2009, from <http://www.istp.murdoch.edu.au/ISTP/casestudies/Case_Studies_Asia/modasia/modasia.html>

Kuhn, A. (2009, April 29). "Concerns Rise with Water of Three Gorges Dam." National Public Radio. Retrieved November 9, 2009, from <http://www.npr.org/templates/story/story.php?storyId=17723829>

Lewyn, M. (2007, September 20). "Debunking Cato: Why Portland Works Better Than the Analysis of its Chief Neo-Libertarian Critic." Congress for the New Urbanism. Retrieved November 9, 2009, from <http://www.cnu.org/node/1533>

Jason, M. and Friedman, D. (2004, March 8). "Hydrogen Fuel Cell Vehicles: Hype or Hero?" Union of Concerned Scientists. Retrieved November 9, 2009, from <http://www.ucsusa.org/clean_vehicles/technologies_and_fuels/hybrid_fuelcell_and_electric_vehicles/hydrogen-fuel-cell-vehicles.html>

"Linking Population, Poverty, and Development: Urbanization, a Majority in Cities." (2007, May). United Nations Population Fund. Retrieved November 9, 2009, from <http://www.unfpa.org/pds/urbanization.htm>

"MAX Light Rail." Wikipedia. Retrieved November 9, 2009, from <http://en.wikipedia.org/wiki/MAX_Light_Rail>

McLaren, W. (2006, May 21). "Chicago Apple Store Has a Green Roof." Treehugger.com. Retrieved November 9, 2009, from <http://www.treehugger.com/files/2006/05/michigan_apple.php>

Melosi, M. V. "The Automobile Shapes the City." Automobile in American Life and Society. Retrieved November 9, 2009, from <http://www.autolife.umd.umich.edu/Environment/E_Casestudy/E_casestudy2.htm>

"Mercury Sources Fact Sheet." (1999). Clean Air Network. Retrieved November 3, 2008, from <http://www.mercurypolicy.org/emissions/documents/hgsources.pdf>

Montagna, J. A. (1981, February 6). "The Industrial Revolution." Yale-New Haven Teachers Institute. Retrieved November 9, 2009, from <http://www.yale.edu/ynhti/curriculum/units/1981/2/81.02.06.x.html>

"Nearly One-Third of Nation's Public Transportation Commuters Live in New York City." (2004, March 2). U.S. Census Bureau News. Retrieved November 9, 2009, from <http://www.census.gov/Press-Release/www/releases/archives/american_community_survey_acs/001701.html>

"New Hybrid Cars—2009-2010 Models." Automobile Magazine. Retrieved November 10, 2009, from <http://www.automobilemag.com/new_cars/27/hybrid_cars/index.html>

"Scientific Assessment Captures Effects of a Changing Climate on Extreme Weather Events in North America." (2008, June 19). National Oceanic and Atmospheric Administration (NOAA) press release. Retrieved November 9, 2009, from <http://www.noaanews.noaa.gov/stories2008/20080619_climatereport.html>

"Nuclear Facts." (2007, February). National Resources Defense Council. Retrieved November 9, 2009, from <http://www.nrdc.org/nuclear/plants/plants.pdf>

O'Neill, H. (2007, October 24). "Greenland's ice sheet melts as temperatures rise." CNN.com. Retrieved November 9, 2009, from <http://www.cnn.com/2007/TECH/science/10/23/greenland.melting/index.html>

"Obama unveils mpg rule, gets broad support." (2009, May 19). MSNBC. Retrieved November 9, 2009, from <http://www.msnbc.msn.com/id/30810514>

"Ocean Planet." (1995). A traveling exhibition of the Smithsonian Institute. Retrieved November 9, 2009, from <http://seawifs.gsfc.nasa.gov/ocean_planet.html>

"Oil: Crude and Petroleum Products Explained—Use of Oil." Energy Information Administration. Retrieved November 9, 2009, from <http://www.eia.doe.gov/neic/infosheets/petroleumproductsconsumption.html>

"Oil price increases since 2003." Wikipedia. Retrieved November 3, 2008, from <http://en.wikipedia.org/wiki/Oil_price_increases_of_2004-2006>

"The Old Times—1930s." Retrieved November 9, 2009, from <http://www.sapiensman.com/old_trains/english.htm>

"Oldest Antarctic Ice Core Reveals Climate History." (2004, June 11). Science Daily. Retrieved November 9, 2009, from <http://www.sciencedaily.com/releases/2004/06/040611080100.htm>

"Our History (Singapore)." (2006). Singapore Urban Redevelopment Authority. Retrieved November 9, 2009, from <http://www.ura.gov.sg/about/ura-history.htm>

"Ozone—Good Up High Bad Nearby." Environmental Protection Agency. Retrieved November 9, 2009, from <http://www.epa.gov/oar/oaqps/gooduphigh/bad.html>#6>

Pantell, S. (2008, February). "Dallas: DART's Light Rail Success Drives Vigorous Expansion Program." Light Rail Now. Retrieved November 9, 2009, from <http://www.lightrailnow.org/news/n_dal_2008-02a.htm>

"Peak Oil Primer." Energy Bulletin . Retrieved November 9, 2009, from <http://www.energybulletin.net/primer.php>

"Pickens Orders 667 GE Turbines." (2008, May 16). Wind Energy News. Retrieved November 9, 2009, from <http://www.windenergynews.com/content/view/1279/43/>

"Power Plants, Pollution, and Soot." (2006, January 9). Environmental Defense Fund. Retrieved November 9, 2009, from <http://www.edf.org/page.cfm?tagID=78>

Proefrock, P. (2007, August 20). "Green Roofs: An Introduction with Pretty Pictures." Ecogeek. Retrieved November 9, 2009, from <http://www.ecogeek.org/content/view/902/>

"Rail Transit Systems in Operation." Light Rail Now. Retrieved December 9, 2008 from <http://www.lightrailnow.org/success1.htm>

"Ratings Highlights." (2008). Greenercars.org. Retrieved November 9, 2008 from <http://www.greenercars.org/highlights.htm>

Roach, J. (2008, April 21). "U.S. Leads World in Wind-Power Growth." National Geographic News. Retrieved November 9, 2009, from <http://news.nationalgeographic.com/news/2008/04/080421-wind-power.html>

Roney, J. M. (2010, September 21). "Solar Cell Production Climbs to Another Record in 2009." Earth Policy Institute. Retrieved October 30, 2010, from <http://www.earth-policy.org/index.php?/indicators/C47/>

Schneider, C. G. and Hill, L. B. (2005, February). "Diesel and Health in America: The Lingering Threat." Clean Air Task Force. Retrieved November 9, 2009, from <http://www.catf.us/publications/view/83>

Shafto, D. "Underground Railroads." Retrieved November 9, 2009, from <http://www.infoplease.com/spot/subway1.html>

"Streetcar History." Portland Streetcar. Retrieved November 9, 2009, from <http://www.portlandstreetcar.org/history.php>

Svoboda, E. with Mika, E. and Berhie, S. (2008, February 8). "America's 50 Greenest Cities." Popsci. Retrieved November 10, 2009, from <http://www.popsci.com/environment/article/2008-02/americas-50-greenest-cities?page=1>

"Table 1-11: Number of U.S. Aircraft, Vehicles, Vessels, and Other Conveyances." RITA Bureau of Transportation Statistics. Retrieved November 9, 2009, from <http://www.bts.gov/publications/national_transportation_statistics/html/table_01_11.html>

"Tax Incentives: SUV Loophole vs. Clean Vehicle Credits Face Uncertain Future." Union of Concerned Scientists. Retrieved November 3, 2008 from <http://www.ucsusa.org/clean_vehicles/solutions/cleaner_cars_pickups_and_suvs/tax-incentives-suv-loophole.html>

"Top 11 Warmest Years On Record Have All Been In Last 13 Years." (2007, December 13). ScienceDaily. Retrieved November 9, 2009, from <http://www.sciencedaily.com/releases/2007/12/071213101419.htm>

"Top 12 CO2-Emitting Countries & Their Per-Capita Emissions (2004)." World Resources Institute. Retrieved November 9, 2009, from <http://www.wri.org/chart/top-12-co2-emitting-countries-their-per-capita-emissions-2004>

"Top 50 Cities in the U.S. by Population and Rank." (2008). Infoplease. Retrieved November 9, 2009, from <http://www.infoplease.com/ipa/A0763098.html>

"Tracing the Surge of Corn-Ethanol Futures Amid USDA Data." (2010, October 11). Seeking Alpha. Retrieved October 30, 2010, from <http://seekingalpha.com/article/229440-tracing-the-surge-of-corn-ethanol-futures-amid-usda-data>

"U.S. Department of Transportation FY2009 Budget." Department of Transportation. Retrieved November 9, 2009, from <http://www.dot.gov/bib2009/2009budgetrequest.htm>

"U.S. Transit Ridership Surges..." (2008, December). Light Rail Now. Retrieved November 9, 2009, from <http://www.lightrailnow.org/news/n_lrt_2008-12a.htm>

"USA: Huge Net Gain for Public Transport in November 2006 Vote." (2006, November). Light Rail Now. Retrieved November 9, 2009, from <http://www.lightrailnow.org/news/n_lrt_2006-11b.htm>

"USFWS & NOAA Deepwater Horizon Response Consolidated Fish and Wildlife Collection Report." (2010, September 15). Retrieved September 15, 2010 from <https://www.piersystem.com/external/content/document/2931/898707/1/ConsolidatedWildlife 15SEP10.pdf>

"Water Science for Schools." United States Geological Survey. Retrieved November 9, 2009, from <http://ga.water.usgs.gov/edu/wuhy.html>

"What is the Greenhouse Effect and Is It Affecting our Planet?" National Oceanic and Atmospheric Administration, National Climatic Data Center. Retrieved November 9, 2009, from <http://lwf.ncdc.noaa.gov/oa/climate/globalwarming.html#q1>

"What's in a Name? Global Warming vs. Climate Change." (2008, December 5). National Atmospheric and Space Administration. Retrieved November 9, 2009, from <http://www.nasa.gov/topics/earth/features/climate_by_any_other_name.html>

"Weekly U.S. Retail Gasoline Prices, Regular Grade." Energy Information Administration. Retrieved November 9, 2008 from <http://www.eia.doe.gov/oil_gas/petroleum/data_publications/wrgp/mogas_home_page.html>

"Wind is a Global Power Source." Global Wind Energy Council. Retrieved November 9, 2009, from <http://www.gwec.net/index.php?id=13>

Wollner, C., Provo, J. and Schablisky, J. (2005, November). "Brief History of Urban Renewal in Portland, Oregon." Retrieved November 9, 2009, from <http://www.pdc.us/pdf/about/portland-ura-history_11-05.pdf>

"World Greenhouse Gas Emissions: 2000, Flow Chart." World Resources Institute. Retrieved November 9, 2009, from <http://www.wri.org/chart/world-greenhouse-gas-emissions-flow-chart>

WEB SITES

American Public Transportation Association
<http://www.apta.com>

Beltline
<http://www.beltline.org>

California Energy Com>mission
<http://www.energyquest.ca.gov/saving_energy/index.html>

Clean Air Task Force
<http://www.catf.us>

Congressional Representatives
<http://www.congress.org/congressorg/home>

EarthSave
<http://www.earthsave.org>

Ecogeek
<http://www.ecogeek.org>

Ecosmart
<http://www.ecosmartconcrete.com>

EIA
<http://www.eia.doe.gov>

Energy Bulletin
<http://www.energybulletin.net>

Energy Information Administration
<http://www.eia.doe.gov>

Environmental Defense Fund
<http://www.edf.org>

Environmental Protection Agency (EPA), Climate Change Web site
<http://epa.gov/climatechange/basicinfo.html>

EPA and U.S. Department of Energy, Energy Star Web site
<http://www.energystar.gov>

Global Development Resource Center
<http://www.gdrc.org/uem/index.html>

Global Warming Newspaper Archive
<http://www.globalwarmingarchive.com>

Global Wind Energy Council
    <http://www.gwec.net>

Government Energy Savers
    <http://www.energysavers.gov>

Green Car Institute
    <http://www.greencars.org>

Infoplease
    <http://www.infoplease.com>

Institute for the Analysis of Global Security
    <http://www.iags.org>

Intergovernmental Panel on Climate Change
    <http://www.ipcc.ch>

Kyoto Protocal
    <http://kyotoprotocol.com>

League of American Bicyclists
    <http://www.bikeleague.org>

Light Rail Now
    <http://www.lightrailnow.org>

Local Governments for Sustainability
    <http://www.iclei.org>

National Aeronautics and Space Administration (NASA)
    <http://www.nasa.gov>

National Geographic
    <http://www.nationalgeographic.com/>

National Geographic News
    <http://news.nationalgeographic.com>

National Oceanic and Atmospheric Administration
    <http://www.noaa.gov>

National Resources Defense Council
    <http://www.nrdc.org>

Pew Center on Global Climate Change
    <http://www.pewclimate.org>

Pew Charitable Trusts (Click Environment/Global Warming)
    <http://www.pewtrusts.org>

Portland Streetcar
    <http://www.portlandstreetcar.org>

Project Vote Smart
    <http://www.votesmart.org/index.htm>

Rails to Trails Conservancy
    <http://www.railtrails.org/index.html>

Renewable Fuels Association
    <http://www.energybulletin.net>

Science Daily
    <http://www.sciencedaily.com>

SourceWatch
    <http://www.sourcewatch.org>

U.S. Department of Energy
    <http://www.energy.gov>

U.S. Department of Transportation
    <http://www.dot.gov>

Union of Concerned Scientists
    <http://www.ucsusa.org>

United Nations Web site on Climate Change
    <http://www.un.org/climatechange>

United States Geological Survey
    <http://www.usgs.gov>

We Campaign
    <http://www.wecansolveit.org>

Wind Energy News
    <http://www.windenergynews.com>

World Resources Institute
    <http://www.wri.org>

Worldwatch Institute
    <http://www.worldwatch.org>

World Wind Energy Association
    < http://www.wwindea.org/home/index.php>

INDEX

ABOUT THE AUTHOR

Sneed B. Collard III is the author of more than thirty award-winning science books including *Teeth, Wings, Reign of the Sea Dragons, Pocket Babies and Other Amazing Marsupials,* and *Science Warriors—The Battle Against Invasive Species.* His book *The Prairie Builders—Reconstructing America's Lost Grasslands* was named the best middle-grade science book of 2005 by Science Books & Films and the American Association for the Advancement of Science. In 2006, he was honored for his writing achievements with the *Washington Post*-Children's Book Guild Nonfiction Award. In 2010, Collard launched his own publishing company, Bucking Horse Books, to focus on quality fiction and nonfiction books for young people. In addition to his writing and publishing activities, Collard is an award-winning speaker, giving presentations to thousands of students and educators every year. Learn more at <www.sneedbcollardiii.com> and <www.buckinghorsebooks.com>.

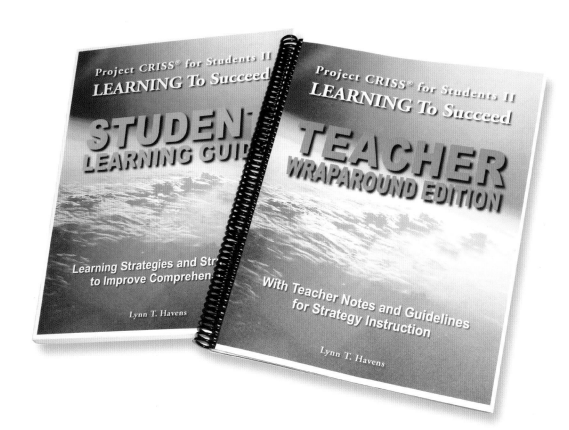

LEARNING STRATEGIES CURRICULUM

Global Warming: A Personal Guide to Causes and Solutions is a companion piece to a learning strategies and content area reading curriculum for secondary students. Developed by Project CRISS® (a professional development provider for educators across the curriculum and throughout the grade levels), this curriculum—comprised of *Global Warming,* student learning guides, and a teacher wraparound edition with PowerPoint® slides and blackline masters—teaches students how to strategically plan and learn in their content area courses. For more information about Project CRISS and to learn more about this curriculum, visit <www.projectcriss.com>.